Legal Studies in International, European and Comparative Criminal Law

Volume 3

The main purpose of this book series is to provide sound analyses of major developments in national, EU and international law and case law, as well as insights into court practice and legislative proposals in the areas concerned. The analyses address a broad readership, such as lawyers and practitioners, while also providing guidance for courts. In terms of scope, the series encompasses four main areas, the first of which concerns international criminal law and especially international case law in relevant criminal law subjects. The second addresses international human rights law with a particular focus on the impact of international jurisprudences on national criminal law and criminal justice systems, as well as their interrelations. In turn the third area focuses on European criminal law and case law. Here, particular weight will be attached to studies on European criminal law conducted from a comparative perspective. The fourth and final area presents surveys of comparative criminal law inside and outside Europe. By combining these various aspects, the series especially highlights research aimed at proposing new legal solutions, while focusing on the new challenges of a European area based on high standards of human rights protection.

As a rule, book proposals are subject to peer review, which is carried out by two members of the editorial board in anonymous form.

More information about this series at http://www.springer.com/series/15393

Francesca Pellegrino

The Just Culture Principles in Aviation Law

Towards a Safety-Oriented Approach

 Springer

Francesca Pellegrino
University of Messina
Messina, Italy

ISSN 2524-8049   ISSN 2524-8057   (electronic)
Legal Studies in International, European and Comparative Criminal Law
ISBN 978-3-030-23177-4   ISBN 978-3-030-23178-1   (eBook)
https://doi.org/10.1007/978-3-030-23178-1

Cover illustration: Maria Isabel Ruggeri

This Springer imprint is published by the registered company Springer Nature Switzerland AG.
The registered company address is: Gewerbestrasse 11, 6330 Cham, Switzerland

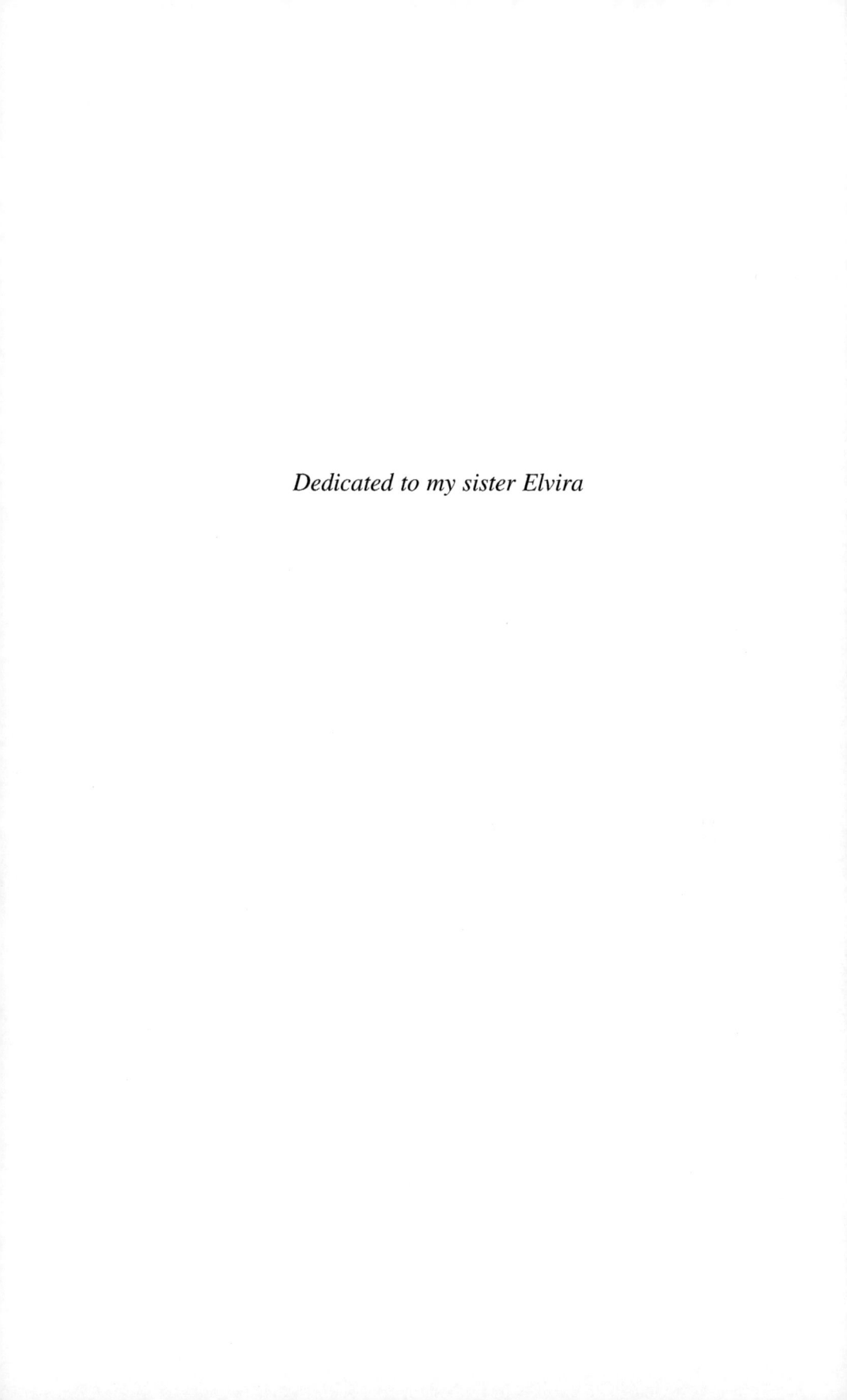

Dedicated to my sister Elvira

Preface

Safety is an essential part of aviation.

Aviation safety means the condition of a system or organization in which risks associated with aviation activities, related to or in direct support of aircraft operations, if not eliminated, are reduced and controlled at an acceptable level.

Measures that improve aviation safety include, in particular, accident investigations, mandatory and voluntary reporting schemes and data analysis programs: they are all instruments of a 'just culture'. This culture recognizes that aviation personnel should freely share and disseminate safety information on deficiencies without fear of punishment.

Therefore, in the last few years, there is an increase of criminal prosecutions against professionals involved in aviation accidents or incidents.

In fact, in the aftermath of several recent disasters, pilots, air traffic controllers or others involved in them have been charged with criminal offenses.

This tendency to criminalize the actions or omissions of professionals depends on the need to find a scapegoat in the face of media pressure and with particular regard to the alleged deterrent nature of penalties. But punishment of pilots or controllers has no demonstrable rehabilitative or deterrent effect.

This trend has increased the fear among aviation professionals that routine operational actions, omission or decisions could become the criminal prosecution ground.

The consequence of this situation is that front-line operators are deterred from reporting deficiencies for fear of being sanctioned for their errors.

Generally, the term 'error' is used in a wider sense, covering both 'gross negligence' and violations. In a technical sense, within the field of aviation safety, human error is an unintentional action or decision, while violations and 'gross negligence' are intentional failures. In turn, there are many types of human errors, intended in a strict sense: slips, lapses, mistakes, which can happen to even the most experienced and well-trained operators.

In the context of a 'just culture', it is certain that prosecution will be limited only to intentional acts which must be suitably sanctioned.

On the contrary, professionals should not be punished for excusable and honest errors that result from ordinary negligence, other than 'gross negligence'.

Nowadays, the challenge in this approach consists of dissuading prosecutors from filing criminal charges against front-line operators for their unintentional acts or omissions.

Messina, Italy   Francesca Pellegrino

Contents

Chapter 1
Origin and Development of a 'Just Culture'

1.1 From Human Error to a 'Safety Culture' and a 'Just Culture'. Reason's Model

According to an ancient, well-known Latin phrase, *errare humanum est,*[1] *i.e.* to err is human. In fact, people make errors, whatever their level of skill, experience or training.[2]

Actually, accidents and incidents form part of our daily life.

When society became industrialized, technology and system complexity[3] generated an increased risk of employee carelessness.[4]

In theory, lapses or weaknesses do not cause a risk to materialize.

In practice, human errors[5] can lead to major accidents[6] or other adverse events. Indeed, human error is considered as a primary contributing factor in disasters and risk events.

The majority of all aviation accidents are attributed to human error,[7] and this is often interpreted as evidence of lack of skill, vigilance or conscientiousness of people at work.

Then we need to ask ourselves a very important question: why does all this happen?

---

[1]Kohn et al. (2000), p. 1 ff., asserts that the problem is not bad people in health care, but it is that good people are working in bad systems that need to be made safer.

[2]Kim (1989), p. 301 ff.

[3]Marti et al. (2011), p. 2 f; Contissa et al. (2012), and Contissa et al. (2013), p. 79 ff.

[4]Pidgeon and O'Leary (2000), p. 15 ff., Reuter (2013), p. 49 ff.

[5]For this concept, see Rasmussen et al. (1987); Shappell and Wiegmann (1997), p. 269 ff.

[6]Anderson (2015).

[7]Johnson (1995), p. 121 ff.

© Springer Nature Switzerland AG 2019
F. Pellegrino, *The Just Culture Principles in Aviation Law*, Legal Studies in International, European and Comparative Criminal Law 3, https://doi.org/10.1007/978-3-030-23178-1_1

There has been a considerable amount of research into the causes of human error,[8] many of these have stressed the physiological influences that shape operator performance.[9]

Especially in a complex system, disasters are due "to the adverse conjunction of a large number of casual factors, each one necessary but singly insufficient to achieve the catastrophic outcome".[10]

In this kind of system, competing task demands and organizational pressures[11] interact with cognitive processes to make people vulnerable to many forms of error.

Therefore, accidents and incidents very often occur as a result of a series of events that lead to disastrous results.

The 'man-machine-environment' theory[12] analyses these three factors capable of causing negative results.

The relationship between man, machine[13] and environment are well known.[14] A 'man-machine-environment' safety system depends not only on the safety training of the staff, but is also closely related to the safety of machinery and equipment and operating locations. In this context, the term environment includes both the physical environment and interpersonal environment. Workplace environment directly affects workers' safety.

---

[8]See Hollnagel (2009). The ETTO principle refers to the fact that people and organisations, as part of their activities, frequently have to make a trade-off between the resources they spend on preparing to do something and the resources they spend on doing it. The trade-off may favour thoroughness over efficiency if safety and quality are the dominant concerns. It follows from the ETTO principle that it is never possible to maximise efficiency and thoroughness at the same time. At this purpose, efficiency means that the level of investment or amount of resources used or needed to achieve a stated goal or objective are kept a slow as possible. Thoroughness means that an activity is carried out only if the individual or organisation is confident that the necessary and sufficient conditions for it exist so that the activity will achieve its objective and not create any unwanted side-effects. See website erikhollnagel.com'.

[9]So Johnson (2005).

[10]Reason (1990b), p. 475 ff.

[11]Hale et al. (1997); Arnesen (1995).

[12]For example, in the USAir Flight 427 disaster, the NTSB concluded that the accident was due to mechanical failure: it determined that the probable cause was a loss of control of the airplane resulting from the movement of the rudder surface to its blowdown limit. USAir Flight 427 was a scheduled flight from Chicago's O'Hare International Airport to Pittsburgh International Airport, with a final destination of West Palm Beach, Florida. On Thursday, September 8, 1994, the Boeing 737 flying this route crashed while approaching runway 28R of Pittsburgh International Airport, located in Pennsylvania. All 132 people on board the aircraft were killed. See Byrne (2002).

[13]For the design of man-machine interface Reason mainly refers to improper allocation of man-machine functions and unreasonable engineering design that can induce man to error easily. See Suchman (1987); Summerton and Benner (2003).

[14]Man-machine-environment system engineering is a new developing synthesis frontier science. It analyses three factors for man, machine and environment of system and the relationship among them to make the system become "safety, cost-effective and highly efficient". So Song and Xie (2014), p. 87 ff. See Yang et al. (2005), pp. 50 ff.; Liu and Zhang (2007), pp. 107 ff.; Piao et al. (2009), p. 44 ff.; Long and Dhillon (2016).

Usually, hazards are prevented from causing human losses by a series of barriers.[15] According to the 'Swiss cheese model'[16]—originally developed by James Reason, a world leading expert on human error, and by Dario Orlandella, both academics of the University of Manchester—each barrier can be represented by a slice of cheese. Each slice has holes in it: they are preconditions, failures that can lead to an accident. This model of accident causation,[17] used in risk analysis and management, demonstrates that, although there are many levels of defense between hazards and accidents, nevertheless often active or latent failures exist in each level and they, if aligned, can allow the disaster to occur. If you can break the chain, you can prevent the accident.

Active failures are errors and violations having immediate negative results and are usually caused by an individual while latent failures refer to less apparent deficiencies in the design of organizational systems or in the environment and are caused by circumstances such as scheduling problems, inadequate training, fatigue, ambiguous communication, organisational influence etc.

Latent failures are usually hidden within an organization until they are triggered by an event likely to have serious consequences, while active failures have an immediate impact on the adverse situation and are usually made by front-line people such as machine operators.

Reason described three levels of human failure that fall within the concept of latent failures.[18] The first refers to the condition that affects front-line operator performance (preconditions for unsafe acts) and involves human biological conditions, such as mental fatigue and stress and also poor communication, coordination practices and frequent interruptions. This failure can lead to errors (skill-based errors, decision errors or perceptual errors)[19] or (routine or exceptional) violations.

'Unsafe supervision' or 'supervisory level' is the second level of human failure. This was identified as an error from lack of sufficient supervision by the organisation's management[20] itself.

---

[15]Hollnagel (2004).

[16]See Reason (1990b), p. 475 ff. See McGrath (2015), model 67, p. 184 f.; Manners-Bell (2014), p. 14 f.; Davina et al. (2016), p. 31.

[17]This model, that has been called 'cumulative act effect' and is considered to be a useful method to understand accident causation, has been subject to criticism.

[18]Reason (1990a), p. 173 f., p. 188 and p. 208 ff.

[19]Human Factors Analysis and Classification System (HFACS), in https://www.skybrary.aero/index.php/Human_Factors_Analysis_and_Classification_System_(HFACS) one can find these clear definitions: "Skill-Based Errors: Errors which occur in the operator's execution of a routine, highly practiced task relating to procedure, training or proficiency and result in an unsafe situation (e.g., fail to prioritise attention, checklist error, negative habit). Decision Errors: Errors which occur when the behaviors or actions of the operators proceed as intended yet the chosen plan proves inadequate to achieve the desired end-state and results in an unsafe situation (e.g, exceeded ability, rule based error, inappropriate procedure). Perceptual Errors: Errors which occur when an operator's sensory input is degraded and a decision is made based upon faulty information".

[20]For more on this topic, see Nordby et al. (1995), p. 33 ff.; Glendon et al. (2006).

Finally, causal factors within the 'organisational level'[21] must be addressed to identify the root cause of any accidents or incidents. They refer to organisational decisions and rules that govern the everyday activities within an organisation and the organisational-level decision making regarding the allocation and maintenance of organisational assets,[22] as well as the working atmosphere within the organisation.

The Reason Model inspired the Human Factors Analysis and Classification (HFACS),[23] a human error framework that was originally used by the US Air Force[24] to investigate and analyze human factor aspects of aviation.

Despite the number of main studies carried out in this field, the traditional culture, aimed at blaming the people involved,[25] without regarding the type of error, is widespread even today. Moreover, in all complex organisational systems,[26] the fault is often not of an individual, but of the system (systemic fault).

However, although errors are involved, they can often be considered 'honest' mistakes that should not be qualified as criminal behaviour.

Therefore, we need to know what a human error is, why people make errors, how many types of errors exist, how errors cause accidents, etc.

Since the 1940s,[27] human factor engineers and psychologists have asked 'what' is responsible for errors, instead of 'who' is responsible.

Contemporary human error theorists catalog errors when examining them.

Several main studies on human errors were carried out by Donald Norman,[28] a well renowned American psychologist and engineer, who differentiated between two types of errors: slips and mistakes.

As Norman stated in his book: "Slips are defined as errors in which the intention is correct, but a failure occurring when carrying out the activities required [...]. Mistakes, by contrast, arise from an incorrect intention, which leads to an incorrect action sequence, although this may be quite consistent with the wrong intention".

Therefore, slips are action errors or execution errors "that are triggered by schemas, a person's knowledge, memories and experiences".[29] They are unintended failures of action or execution that occur almost every day because attention is not

---

[21] See Caldwell (2018), Chapter 2, in particular para. 2.3.2.

[22] Cromie et al. (2013).

[23] "The Human Factors Analysis and Classification System (HFACS)" Approach, July-August 2004, developed by Scott Shappell and Doug Wiegmann. See Shappell (2000), Wiegmann and Shappell (2003), Shappell and Wiegmann (2001), p. 59 ff. See also Leplat (1982) and Ford et al. (1999), p. 343.

[24] HFACS Analysis of Military and Civilian Aviation Accidents: A North American Comparison, ISASI, 2004. See also Li and Harris (2005), p. 67 ff.

[25] In contrast, the United States does not criminalize operators involved in controversial circumstances after aviation accident, but the penalties are primarily financial. See Lintner and Dunlap (2013), p. 52 ff.

[26] Scholtze (2014), para. 1.

[27] Festinger (1957); Johnston et al. (1994).

[28] Norman (1988, 2013).

[29] So Strauch (2017), p. 19.

fully applied to the task in hand. Mistakes are errors of thought in which a person's cognitive activities lead to actions or decisions that are contrary to what was intended. They are errors in choosing an objective or specifying a method of achieving it, whereas slips are errors in carrying out an intended method for reaching an objective.

According to Sternberg's opinion,[30] "slips are most likely to occur (a) when we must deviate from a routine, and automatic processes inappropriately override intentional, controlled processes; or (b) when automatic processes are interrupted - usually as a result of external events or data, but sometimes as a result of internal events, such as highly distracting thoughts".

Jens Rasmussen,[31] a system safety[32] and human factors professor at the University of Risø (Denmark), expanded the classification that Norman described, defining three different types of errors made by the operators, associated to three types of operator performance: 'skill-based', 'rule-based' and 'knowledge-based'.

In his view, the operator acts at one of these levels depending on the nature of the task and the level of experience required in the particular situation.

He defined the skill-based behaviour as a 'sensory-motor performance during acts or activities which, following a statement of an intention, takes place without conscious control as smooth, automated, and highly integrated patterns of behavior'.[33]

Therefore, skill-based performances are routine, highly practiced tasks that we do in almost automatic fashion. Any human error that occurs at this level is most often due to attentional lapses.

A rule-based error is more advanced than skill-based. It applies rules to situations that are similar to those that operators have encountered through experience and training. Therefore, it results from the inability to recognize or understand the situations or circumstances encountered. Errors made at this level often result from a misclassification of the situation and a consequent application of the wrong rule.

Rasmussen believed that the highest level of performance is knowledge-based. Knowledge-based performance takes effect when, rather than applying simple tasks and rules to situations that are similar to those previously encountered, the operator applies information formerly learnt or obtained through past experiences.[34] In this case, errors result from shortcomings in the operator's knowledge or limitations in his ability to apply existing knowledge to new situations.[35]

These and other main studies created the basis for the elaboration of 'just culture' principles.

---

[30] Sternberg (1996). Robert Sternberg is an American psychologist and psychometrician. He is Professor of Human Development at Cornell University.

[31] Jen Rasmussen was senior member of IEEE, institute of Electrical and Electronic Engineers, New York. See Rasmussen (1982, 1983), p. 257 ff.

[32] McDonald et al. (2002), p. 205 ff.; Catino (2013).

[33] Rasmussen (1983), p. 258.

[34] *Ibid.*, p. 259.

[35] See also Wickens et al. (1998).

In the 1990s, James Reason defined error[36] as "a generic term to encompass all those occasions in which a planned sequence of mental or physical activities fails to achieve its intended outcome, and when these failures cannot be attributed to the intervention of some chance agency".[37]

He stressed that human errors can be addressed in two different ways: the person approach and the system approach. The first one concentrates on the errors of individuals, blaming them for inattention, moral weakness or forgetfulness. According to this approach, punishing someone's misbehaviour encourages conformity to a rule and urges people to take precautions to reduce the risks accompanying inadvertent behaviour. The second approach focuses on the conditions under which individuals work and tries to find out if obstacles had been in place to avoid errors or mitigate their effects.

The first scenario is attributable to the so-called 'blame (or blaming) culture", while the second one is referred to as the 'just culture' or 'no blame culture'.

The two equivalent expressions 'no blame culture' and 'just culture' have been used by Reason in his well-known book "Managing the Risks of Organizational Accidents",[38] in which "engineering a 'just culture '" is considered the most important early step that aims to overcome the traditional and wide-spread 'blaming culture'.[39]

Looking at the events in detail, Reason believes that all human actions involve three basic elements: an intention that specifies an immediate goal of a future behaviour; the actions triggered by this intention and the consequences of these actions, which can or cannot achieve the desired objective.

In Reason's opinion, it should be considered that only a very small proportion of unsafe human actions are deliberate and a large part of them consists of honest errors (*e.g.* slips, lapses, mistakes etc.). Lapses[40] are actions resulting from plain forgetfulness and occur more frequently in maintenance or installation procedures, where an omission of a single task can be critical.

In addition, Reason sub-divided mistakes into two categories: 'knowledge-based mistakes' and 'rule-based mistakes'. The first ones are the kinds of errors that would occur during the failure to correctly appreciate a situation or make the appropriate

---

[36]Senders and Moray (1991), p. 25, defined error as "something that has been done which was not intended by the actor, not desired by a set of rules or an external observer, or that led the task or system outside its acceptable limits". See also Woods et al. (1994), p. 2, defined error as "a specific variety of human performance that is so clearly and significantly substandard and flawed when viewed in retrospect that there is no doubt that should have been viewed by the practitioner as substandard at the time the act was committed or omitted". More recently, Strauch (2004), p. 21 defined error as "an action or decision that results in one or more unintended negative outcomes".

[37]Reason (1990a), p. 5.

[38]Reason (1997). See also Reason (2000a), p. 768 ff.

[39]See Pepe and Cataldo (2011), p. 56 f.

[40]Otherwise, Wickens et al. (2016) say that "whereas slips represent the commission of an incorrect action, different from the intended one, lapses represent the failure to carry out an action at all" (see chapter 7, par. 5.1).

decision.[41] It is due to the lack of knowledge or expertise[42] to understand the situation. 'Rule-based mistakes' are made with confidence, but by choosing a different procedure or a wrong action because of a lack of situational assessment.

These two categories of mistakes do not differ from those of Rasmussen's model.

He believes that it is unfair to indiscriminately sanction every possible unsafe conduct following a 'blame culture' approach.

If the large majority of unsafe acts are 'honest errors', it means that they should be reported without fear of sanction. To this end, a crucial trust is required.

However, a "no-blame" culture *per se* is "neither feasible nor desirable", because it would leave unpunished deliberate behaviour, such as 'willful misconduct'. In addition, blanket amnesty on all unsafe acts would cause a lack of credibility in the public eye.

In his view, "a safety culture depends upon first negotiating where the line should be drawn between unacceptable behaviour and blameless unsafe acts".[43]

It is true that the civil aviation system should promote a 'safety culture'[44] facilitating the spontaneous reporting of occurrences and thereby disseminating an informed culture, but therefore an effective reporting culture depends on how the single organization handles and balances blame and punishment.[45]

In other words, an effective reporting culture is underpinned by a 'just culture', in which the culpability[46] line is more clearly drawn.[47]

Synthetically, according to the Reason model,[48] a 'safety culture' has the following characteristics:

(a) it is an 'informed culture'[49] in which those who manage a complex system or operate in it have detailed and current knowledge about technical, organizational, environmental and human factors that determine the entire safety system;
(b) it is a 'reporting culture' in which people are willing and prepared to report their errors and near-misses, in an atmosphere where they have confidence to report safety concerns, without fear of blame;

---

[41]Woods (2017).

[42]See Dismukes et al. (2017), and Flin et al. (2008).

[43]GAIN (Working Group E. Flight Ops/ATC Ops Safety Information Sharing) (2004), A Roadmap *to a* Just Culture: Enhancing the Safety Environment, 1st edition, vi.

[44]The term was first applied in 1986, with respect to the Chernobyl disaster. See Carnino and Weimann (1995); Wiegmann et al. (2002).

[45]See Johnson (1995), p. 15, p. 33 ff., Pellegrino (2004), p. 83 ff.

[46]Cromie and Bott (2016), p. 258 ff.

[47]Reason (1998), p. 293 ff. See also Hudson (2003a), p. 7 ff. and Hudson (2003b), p. 27 ff.

[48]See also Reason (2000b), p. 3 ff. See also Pidgeon and O'Leary (1994), p. 21 ff., where they suggested that a good safety culture's concern for safety is distributed and endorsed throughout the organisation; Pidgeon (1998), vol. 12, issue 3, p. 202 ff.; Guldenmund (2000), p. 216 ff.; Cooper (2000), pp. 111 ff.; Wiegmann et al. (2002); Harris and Li (2006), p. 345 ff.; Morley and Harris (2006), p. 3 ff.; Cooper and Findley (2013).

[49]Ferguson and Fakelmann (2005), p. 33 ff.

(c) it is a 'learning culture'[50] where an organization whose workers are able to learn from its mistakes, having the willingness and the competence to draw the right conclusions from its safety information system and the will to implement major reforms;
(d) it is a 'flexible culture' in which an organization is able to reconfigure itself in the face of certain kinds of dangers, "often shifting from the conventional hierarchical mode to a flatter mode";
(e) it is a 'corporate culture' intended as 'the set of unwritten rules that govern acceptable behaviour within and outside the organisation. It emanates from the strategic apex of the company and colours all of its activities'[51];
(f) it is a *just culture*, a 'no blame' culture, characterized from an atmosphere of trust, in which people are encouraged (or even rewarded) for providing essential safety-related information, but in which they are also clear about where the line between acceptable and unacceptable behaviour must be drawn.[52] Therefore, 'just culture' requires a set of good practices and safety related attitudes[53] in high-risk workplaces.

Therefore, to create or maintain such a safety-conscious and trusted aviation atmosphere, the introduction and dissemination of a 'just culture' is needed, described by Reason as "an atmosphere of trust in which people are encouraged (even rewarded) for providing essential safety-related information but in which they are also clear about where the line must be drawn between acceptable and unacceptable behaviour".[54]

1.2 Just Culture' in Aviation

1.2.1 Application of 'Just Culture' Principles to the Aviation Sector

The reasons that explain the fact that 'just culture' principles have initially taken root in aviation depend on the characteristics of this sector.

In fact, the aviation industry is a very complex system, characterized by high technology and high risk activities, with greater potential for catastrophic consequences (accidents).

---

[50]Rasmussen (1983). See also Catino and Patriotta (2013), p. 437 ff.

[51]Pidgeon and O'Leary (1994), p. 21 ff.; Daniels (2017).

[52]For a commentary of the Reason model, see Hudson (2001), p. 8 and Hudson (2003b), p. 27 ff.; Sumwalt (2007), p. 38.

[53]Jeffries (2011), p. 202 f.; Pellegrino (2014, 2015).

[54]Reason (1995), p. 1708 ff. See also Harris (2011), p. 292.

In aviation, an 'accident' is defined by the Annex 13[55] to the Chicago Convention on International Civil Aviation[56] as an occurrence associated with the operation of an aircraft, which takes place from the time any person boards the aircraft with the intention of flight until all such persons have disembarked, and in which (a) a person is fatally or seriously injured, (b) the aircraft sustains significant damage or structural failure, or (c) the aircraft goes missing or becomes completely inaccessible.

The same Annex defines an 'incident' as an occurrence, other than an accident, associated with the operation of an aircraft, "that affects or could affect the safety of operation".[57]

Finally, a 'serious incident' is defined as an incident, associated with the operation of an aircraft, "involving circumstances indicating that an accident nearly occurred".[58] This is equivalent of a near miss. Therefore, the difference between accident and serious incident lies only in the result.[59]

In this context, traditionally front-line personnel have been considered as solely responsible for accidents, serious incidents or incidents.

However, operators do not act in isolation, but as part of an organization. Their actions reflect processes, procedures and attitudes (e.g. the 'safety culture') of the other part of the organization.

Accordingly, safety culture can be regarded as "beliefs, values, attitudes, behaviours, etc., related to safety and shared by an organization at large (culture) or smaller groups within an organization (subcultures)".[60]

Therefore, managing safety in complex systems like aviation requires a systems approach that takes a deeper view, looking for organizational deficiencies that can lead to an accident.

In this context, it is very interesting to note that active failures are typically carried out by operational personnel, while latent failures are due to decisions or actions typically performed by managers.[61]

---

[55]See Annex 13, *Aircraft Accident and Incident Investigation*, eleventh edition, July 2016, chapter 1. For a commentary, see Pellegrino (2003), p. 805 ff. and Pellegrino (2007), p. 107 ff.; Beveridge (2009), p. 151 ff.; Challinor (2016).

[56]Convention on International Civil Aviation, done at Chicago on the 7th of December 1944. For a commentary, see Giannini (1944), p. 101 ff.; Giannini (1953a), p. 11 ff., Giannini (1946), p. 83 ff., Giannini (1949), p. 41 ff., Giannini (1952), p. 729 ff., Giannini (1953b), p. 269 ff., Giannini (1953c), p. 130 ff.; Giannini (1953c), p. 101 ff.; Quercietto (1954), p. 1225 ff.; Fragali (1967), p. 159 ff.; Mateesco Matte (1995), p. 641 ff.; Lu (2013), p. 13 ff.; Haanappel (2003), p. 43 ff.; Milde (2008), p. 17 ff. and 63 ff.; Benkö (2012); Havel and Sanchez (2014), p. 28 ff.; Weber (2017), chapter 2.

[57]For a commentary of these definitions, see Pellegrino (2007), p. 27 ff.

[58]Annex 13, chapter 1.

[59]Examples of serious incidents can be found in Attachment C of Annex 13 and in the Accident/Incident Reporting Manual (Doc 9156).

[60]These considerations are drawn from Perez Gonzales (2012), p. 13 ff. See also Khatib (1996), p. 8 ff., p. 19; Patankar et al. (2012), p. 179 ff.

[61]See Perez Gonzales (2012), 13 ff., chapters 2 and 3, where the following is stated: "Active failures are typically carried out by operational personnel, whereas latent failures are due to decisions or

Since the 2000s, Eurocontrol (European Intergovernmental Organisation for the Safety of Air Navigation)[62] has played a crucial role in the implementation process of the concept of 'just culture', which was seen as a key for further improvement of aviation safety.

Over the last few years, this Organisation has in fact promoted significant developments in this crucial safety area with a devoted Just Culture Task Force (JCTF).[63]

JCTF's work started in 2007 with the aim "to promote, debate and discuss the legal questions relating to safety and justice, to foster and support dialogue between safety and judicial experts, and to develop relevant guidance material and policies in order to support the implementation and dissemination of a Just Culture concept".[64]

The first document drawn up by the above Organisation on this subject is entitled "Just Culture Guidance Material for Interfacing with the Judicial System"[65] and was elaborated with the aim of identifying the conditions for creating this culture at national and international levels.

In fact, this document stresses that experience has shown that we can learn from honest mistakes and information sharing about them, but on condition that they are freely reported by front-line operators, without fear of negative consequences. The key words for implementing 'just culture' principles are openness and information sharing.

According to Eurocontrol's view, there are two main questions that require answers: 'Who draws the line between acceptable and unacceptable behaviour?' and 'How is safety data protected?'

In answer to the first question, it is desirable to report the exact words that were written by Sidney Dekker:

---

actions typically performed by decision-makers such as managers and system designers" [. . .] "Safety feedback systems are also required for reporting active risks and hazards, auditing the organization to uncover latent failures and dissemination of information throughout the industry". See Douglas (1992).

[62]International organization established by Convention relating to Co-operation for the Safety of Air Navigation of 13 December 1960, as amended by the amending Protocol of 12 February 1981 (*amended Convention*) and by the Brussels Protocol of 27 June 1997 (revised Convention), consolidating the Eurocontrol International Convention relating to Co-operation for the Safety of Air Navigation of 13 December 1960; Sardella (1995), p. 12 ff.; Schwenk and Schwenk (1998), p. 129; Spadoni (2001), p. 259 ff.; Trovò (2008), p. 163 ff.

[63]The Task Force is composed of representatives of the European Commission, the European Aviation Safety Agency (EASA), the International Federation of Air Traffic Controllers' Associations (IFATCA), the European Regional Airlines Organisation (ERA), the European Cockpit Association (ECA), accident investigation boards, national prosecutors, Civil Aviation Authorities (CAAs), Air Navigation Service Providers (ANSPs), the National Transportation Safety Board (US NTSB) and ICAO.

[64]So https://www.eurocontrol.int/articles/just-culture.

[65]Eurocontrol (2008), Just Culture Guidance Material for Interfacing with the Judicial System, Edition 1, EATM (European Air Traffic Management) Infocentre, Bruxelles, 11 February 2008 (Reference: 08/02/06-07). See North (2000), p. 66.

Culpability is socially constructed: the result of deploying one language to describe an incident, and of enacting particular post-conditions. Different accounts of the same incident are always possible (*e.g.* educational, organizational, political). They generate different repertoires of countermeasures and can be more constructive for safety. The issue is not to exonerate individual practitioners but rather what kind of accountability promotes justice and safety: backward-looking and retributive, or forward-looking and change-oriented.[66]

This implementation process has continued thanks to another Eurocontrol paper, entitled "Model Policy regarding criminal investigation and prosecution of civil aviation incidents and accidents",[67] that identifies a condition for establishing a 'just culture' at national level, *i.e.* enhanced cooperation and coordination between safety and judiciary. This necessary cooperation is required between national prosecuting authorities of Eurocontrol Member States[68] and safety authorities by means of agreements.

The stated goal of this policy is to provide guidance regarding criminal investigation and prosecution of potential criminal charges resulting from aviation accidents or from incidents reported under mandatory or voluntary occurrence reporting schemes, subject to the attention of prosecutors.

Finally, both Eurocontrol and ICAO issued many documents to answer the question: How is safety data protected in order to ensure a 'just culture'? They will be mentioned in the following section.

One among them deserves to be singled out, namely Annex 19[69] on Safety Management, as amended in 2016[70] with the aim of ensuring appropriate use and protection of safety information and to support proactive safety improvement strategies.

In this context, definitions for safety data and safety information have also been developed to provide clarity to the relative provisions. In particular, 'safety data' is defined as "a set of facts or a set of safety values collected from various aviation-related sources, which is used to maintain or improve safety", while 'safety

---

[66]Dekker (2008).

[67]The Provisional Council of the Eurocontrol Organisation has unanimously endorsed this Policy at its session of 10th May 2012. See Eurocontrol (ed), Just Culture Policy, Bruxelles, September 2012, which presents the said Model Policy.

[68]Eurocontrol is an intergovernmental organisation with 41 Members: Albania, Armenia, Austria, Belgium, Bosnia and Herzegovina, Bulgaria, Cyprus, Croatia, Denmark, Estonia, Finland, France, Georgia, Germany, Greece, Ireland, Italy, Latvia, Lithuania, Luxembourg, Macedonia, Malta, Moldova, Monaco, Montenegro, Norway, Benelux countries, Poland, Portugal, United Kingdom, Czech Republic, Romania, Serbia, Slovakia, Slovenia, Spain, Sweden, Switzerland, Turkey, Ukraine, Hungary.

[69]On this subject see ICAO Annex 19 (Safety Management. The first edition of Annex 19 was February 2013 and becomes applicable on 14 November 2013. The Safety Management Annex 19, representing the first new ICAO Annex to be adopted in over 30 years, became applicable on 14 November 2013. For a commentary see Rallo (2014), p. 101 ff.

[70]Annex 19 (Safety Management), Second Edition, was adopted by the Council on 2 March 2016, became effective on 11 July 2016 and will be applicable on 7 November 2019. This edition supersedes the previous edition 2013.

information' is defined as "safety data processed, organized or analysed in a given context so as to make it useful for safety management purposes".

A note has been added to the definition of safety data, which specifies that such data are collected from proactive or reactive safety-related activities. Some of these include: accident and incident investigations; safety reporting; continuing airworthiness reporting; operational performance monitoring; inspections, audits, surveys; safety studies and reviews.

Another important new addition is that the protection of safety data and safety information in voluntary reporting systems is carried out through an 'international standard' in order to ensure their continued availability and greater uniformity among States. The protection of safety data and safety information in mandatory reporting systems, which differ among different legal systems, is agreed upon a recommendation or 'recommended practice'.[71]

It should be pointed out that a standard is defined by ICAO as any specification "the uniform application of which is recognized as necessary for the safety or regularity of international air navigation and to which Contracting States will conform, in accordance with the Convention"[72] (*i.e.* a binding rule), whereas a recommended practice is defined as any specification "the uniform application of which is recognized as desirable in the interest of safety, regularity or efficiency of international air navigation and to which Contracting States will endeavor to conform, in accordance with the Convention"[73] (*i.e.* non-binding rule).

1.2.2   Relationship Between 'Just Culture', Safety Investigation and Occurrences Reporting

There is a need to draw lessons from safety investigation of accidents and incidents[74] so as to take appropriate action to prevent the repetition of such events. Technical incident investigations take place with the intention of enabling improvement of equipment, operational procedures and training.

It is necessary and mandatory to carry out safety investigations of accidents and serious incidents,[75] which are the most severe events. Furthermore, it is desirable that less serious occurrences be investigated in order to learn lessons and take

---

[71]For a summary of the new provisions contained in Annex 119, Second Edition 2016 see COSCAP (Cooperation Development of Operational Safety and Continuing Airworthiness Programme), 16th Steering Committee Meeting, Update on ICAO State Safety Programme and Safety Management System Implementation, Discussion Paper 6 (presented by the ICAO Montrea), Summary, para. 2.2.

[72]Assembly Resolution A36-13, Appendix A. ICAO Doc 9902, Assembly Resolution in force as of 28 September 2007.

[73]*Ibid.*

[74]See Czech et al. (2014).

[75]According to Article 5 of Regulation (EU) No. 996/2010 on the investigation and prevention of accidents and incidents in civil aviation, examined in the following. Article 5(1) of ICAO Annex

preventive, proactive,[76] and precautionary measures, thus preventing that these occurrences turn into more serious effects.

Experience has shown that we can learn so much from honest mistakes. Spontaneous reporting is considered to be a means of ensuring the implementation of 'just culture' principles, in an atmosphere of complete trust.

This implementation is now seen as key for further improvement of aviation safety, through more detailed investigation and better reporting of aviation occurrences.

In accordance with the above mentioned Reason model,[77] which considers 'safety culture' as an informed culture, a variety of safety analyses, investigations, reporting systems and other preventive measures and initiatives have to be implemented in the aviation industry as a way to enhance operational safety.

Accidents and incidents are investigated in order to identify the causes of these adverse events.

The safety investigation aims at preventing future accidents and incidents (preventive measures),[78] while the judicial investigation focuses on identifying and attributing blame and responsibility (reactive measures).

In this respect, Article 3(1) of ICAO Annex 13 clarifies 'the sole objective of the investigation of an accident or incident shall be the prevention of accidents and incidents. It is not the purpose of this activity to apportion blame or liability'.

In addition, Article 3, point *e)*, of the Council Directive 94/56/EC,[79] in no uncertain terms, defined 'investigation' as "a process conducted for the purpose of accident and incident prevention which includes the gathering and analysis of information, the drawing of conclusions, including the determination of cause(s) and, when appropriate, the making of safety recommendations".[80]

The preventive nature of a technical investigation was indirectly repeated in Article 4 of the same Directive as follows: "The investigations referred to in paragraph 1 shall in no case be concerned with apportioning blame or liability", thus underscoring the distinction between judicial inquiry (which aims at the ascertainment of facts and, in case of conviction, the punishment of the guilty) and technical investigation (which aims at the prevention of similar fatal or adverse events in the future).

Directive 94/56/EC has been repealed by the above mentioned Regulation No. 996/2010, which has similarly defined the safety investigation as

---

13, mentioned above, imposes to carry out an investigation only of accidents, while recommends to carry out an investigation of serious incidents.

[76]Dusi (2017), p. 35 ff.

[77]Reason (1997), chapter 9.

[78]Hobbs and Robertson (1995), p. 468 ff.

[79]Council Directive 94/56/EC of 21 November 1994 establishing the fundamental principles governing the investigation of civil aviation accidents and incidents (O.J. L 319 of 12 December 1994, p. 14).

[80]Comenale Pinto (2015), p. 91 ff.

a process conducted by a safety investigation authority for the purpose of accident and incident prevention which includes the gathering and analysis of information, the drawing of conclusions, including the determination of cause(s) and/or contributing factors[81] and, when appropriate, the making of safety recommendations (Article 2(14)).

In addition, Article 1 of this Regulation stresses that it

aims to improve aviation safety by ensuring a high level of efficiency, expediency, and quality of European civil aviation safety investigations, the sole objective of which is the prevention of future accidents and incidents without apportioning blame or liability, including through the establishment of a European Network of Civil Aviation Safety Investigation Authorities.

More recently, according to the IATA (International Civil Transport Association) opinion, "The sole aim of the safety investigation is to find out what went wrong and to use this information to prevent a similar accident from happening again. The criminal investigation, on the other hand, tries to find out who is to blame for an accident and then punish those concerned".[82]

The latter aims to apportion blame or liability.

Therefore, the only purpose of a safety investigation is the prevention of future accidents. It is conducted with the sole intention of identifying the causes of the accident and making safety recommendation to prevent the recurrence of similar events.

In other words, the aim of technical investigations is to determine the causes of an accident or serious incident in order to avoid similar events in the future and to increase safety standards, by following a preventive approach, while judicial inquiries are aimed at prosecuting people found guilty or those believed to be responsible, by adopting a repressive approach.

In the light of all these provisions, there is no doubt that the purpose of an investigation into an aircraft accident or incident is to prevent their re-occurrence.

In contrast to this purpose, criminal-law standards are generally determined by society's intolerance of certain acts or omissions, which are judged by society to be unacceptable. Some form of punishment is often imposed upon as a deterrent or sometimes to ensure justice or occasionally as simple retribution.[83]

This leads to another observation. Technical investigation follows an *ex post* approach, by acting after the adverse event happens, while safety has to be proactive rather than wait for the incident to occur and then look for suitable safety measures (*i.e.* reactive).

From a proactive perspective, Directive 2003/42/EC[84] on occurrence reporting in civil aviation recognized that the 'sole objective' of occurrence reporting is

---

[81]Haslam et al. (2004), p. 401 ff.

[82]IATA, Safety: The Blame Game, Airlines, 1 June 2012. See also IATA, Operational Safety Audit (IOSA), Montreal (Canada), 2011.

[83]So Lawrenson and Braithwaite (2018), p. 251 ff. and, in particular, para. 3.

[84]EC, Parliament and Council Directive 2003/42/EC of 13 June 2003 on occurrence reporting in civil aviation (O.J. L 167 04/07/2003, p. 23). For a commentary, see Pellegrino (2007), p. 273 ff.; Ingratoci and Michaelides-Mateou and Mateou (2014), p. 265 ff.

"the prevention of accidents and incidents and not to attribute blame or liability" (Article 1).

Regarding relations between judicial and technical investigation, the same Regulation No. 996/2010, in recital 23, clarifies that "an accident raises a number of different public interests such as the prevention of future accidents and the proper administration of justice" [...] "the right balance among all interests is necessary to guarantee the overall public interest".

These interests[85] are the needs of the aviation industry and the objectives of the judiciary and they go beyond the individual interests of the parties involved and beyond the specific event.

In other words, the well-recognized public interests are: the right of the highest possible level of safety (through accident investigation and reporting systems) and the right of independent and impartial administration of justice[86] and freedom of public information through appropriate constitutional and legislative provisions.[87]

Therefore, aviation safety and the administration of justice are two different realities.

Aviation safety is by nature international, dynamic, pro-active, forward-looking and obviously safety-oriented,[88] whereas the administration of justice is by nature national, static and resistant to progressive change and guilt-responsibility-compensation-oriented.[89] Safety is a highly complex, socio-technical system, while judiciary is accepted by society and very sensitive to the rule of law. The goal of safety is to gain information and measures needed to increase it, whereas judiciary aims to determine blame and responsibility and adopt corresponding punitive measures. The methodology is also different: safety is based on systemic investigation in order to reconstruct the event in the broadest possible manner from the perspective of the involved operators, while the administration of justice is based on outcome-based investigation with hindsight, blaming a single individual.

'Just culture' is somewhere in between, as it aims to balance and satisfy the different interests of two unique and basically incompatible worlds.

Notwithstanding these differences, Regulation No. 996/2010 stresses that a balance between these divergent objectives is required.

In particular, recital 23 emphasises that both aviation safety and the administration of justice would benefit from an established equilibrium,[90] ensured through

---

[85]See Balcerzak (2017), p. 5 ff.

[86]Hollnagel (2013), p. 10 f.

[87]See Working Paper of ICAO Assembly, 36th Session, Technical Commission, A36-WP/232, TE/76, 18 September 2007, p. 2 "Protection of certain accident and incident records and of safety data collection and processing systems in order to improve aviation safety", in Organization ICAO, Montréal. See Milde (2012), p. 159.

[88]So Bijlsma (2013), p. 64 ff.

[89]So Information Paper of Safety Information Protection Task Force (SIPTF), fourth Meeting, Montréal, 21–25 January 2013, "Just Culture", SIPTF/4-IP/2 of 17 January 2013, presented by Professor Roderick van Dam, 1.6.

[90]Weston (2014), p. 69 ff.

application of the principle of 'just culture', in spite of their different purposes. On the contrary, the absence of a 'just culture' in an organization affects both safety and justice at the same time.

According to subsequent recital 24, "The civil aviation system should equally promote a non-punitive environment facilitating the spontaneous reporting of occurrences and thereby advancing the principle of 'just culture'".

Therefore, spontaneous reporting is considered to be a means of ensuring the application of the principle of 'just culture' in an atmosphere of complete trust, without the fear of being punished.[91]

Under Article 8(4) of the repealed Directive 2003/42/EC, Member States had to refrain from instituting proceedings[92] in respect to: "unpremeditated or inadvertent infringements of the law which come to their attention only because they have been reported under the national mandatory occurrence-reporting scheme" except in situations of 'gross negligence'.

This provision, which showed the need to refrain from punishing people in respect to unpremeditated and inadvertent infringements, in line with principles of 'just culture', has been repeated by Article 16(6) of Regulation (EU) No. 376/2014,[93] which is still in force and has been integrated by Commission Implementing Regulation (EU) 2015/1018.[94]

In this context, recital 36 of the same Regulation stresses that the civil aviation system should promote a 'safety culture', by facilitating the spontaneous reporting of occurrences and "thereby advancing the principle of a just culture".

---

[91]So Pellegrino (2013), p. 472; Hollnagel (2013), p. 81 ff.

[92]See Panelli (2015), p. 231 ff.

[93]Regulation (EU) No. 376/2014 of the European Parliament and of the Council of 3 April 2014 on the reporting, analysis and follow-up of occurrences in civil aviation, amending Regulation (EU) No. 996/2010 of the European Parliament and of the Council and repealing Directive 2003/42/EC of the European Parliament and of the Council and Commission Regulations (EC) No. 1321/2007 and (EC) No. 1330/2007 (O.J. L 122 of 24 April 2014, 18).

[94]Commission Implementing Regulation (EU) 2015/1018 of 29 June 2015 laying down a list classifying occurrences in civil aviation to be mandatorily reported according to Regulation (EU) No. 376/2014 of the European Parliament and of the Council (O.J. L 163 of 30 June 2015).

1.3 'Just Culture' in Healthcare

1.3.1 Application of 'Just Culture' Principles to the Healthcare Sector

The principle of 'just culture', initially conceived in the world of aviation, has been subsequently recognized in many safety-critical branches, such as other modes of transport,[95] medicine,[96] nuclear technology,[97] high-risk industries[98] and so on.

Historically, a 'blame culture' has been the norm in health-care field, in which individual practitioners were held accountable for any, albeit slight, error.

Nevertheless, as a result of a typical movement in the aviation industry, in the late 1990s, David Marx,[99] an expert on national and international health-care quality and safety management, opened the new frontier of a 'just culture', by applying this concept to medical risks,[100] considering the possible applicability of these principles to all complex systems.[101]

In Marx's view, the single greatest impediment to safety prevention is that we punish people for making mistakes,[102] according to a 'blame culture' approach. Actually, an honest mistake should not be punished in order to encourage reporting of failures or deficiencies and information sharing from the personnel involved.

In his opinion, the 'just culture model', unlike the punitive approach, seeks to create a trust-based environment that encourages health-care personnel to report potential and actual errors for the purpose of a proactive investigation system.[103]

For this personnel, safety is an individual responsibility commensurate with training, experience and professional standards. Notwithstanding the utmost

---

[95] Amalberti (2001), p. 110 ff.

[96] Pellegrino (2004), p. 83 ff.; Berlinger (2005), Merry (2009), p. 265 ff.

[97] Kennedy and Kirwan (1995), p. 282 ff; Westrum (1996).

[98] Heinrich et al. (1980), and Perrow (1984).

[99] David Marx, President of Outcome Engineering, a risk management company located in Dallas, USA. See Marx (1997), pp. 1 ff. and Marx (2009), in which he explores the role of human error in society from aviation and health care and speaks about culpability, accountability, behavioural choices, system design and the 'benefits' of a 'just culture'. He says: "The challenge lies in distinguishing between a system that might create risks, human error which may result in a bad outcome, and reckless behavior that intentionally puts lives or organizations at risk". See Outcome Engineering LLC. (2008) Just Culture Training for Healthcare Managers. See also Weston (2014), p. 69 ff.

[100] Hudson (2003a), p. 7 ff. See also Perrow (1984), and Green (2003).

[101] Westrum (1993).

[102] Merry and McCall Smith (2001), and Morreim (2004), p. 213. See also Leape (1994), p. 1851 ff. He says that in a punitive stem, no one learns from their mistakes. More recently, see Dekker (2007), chapter 5.

[103] So Marx (2001), p. 25. See also Whitehead et al. (2017), p. 126; Barnsteiner and Disch (2012a), p. 407 ff.

importance of safety in the workplace, how can a culture be established which encourages open reporting of adverse events and risky situations?

According to Marx's opinion, medical organisations should promote safety reporting[104] and 'just culture' principles, being responsible for encouraging positive safety behaviours by providing the proper environment,[105] training, tools and procedures[106] to their staff. All should cooperate to build trust across the organization.

In his opinion, a health care system is obliged to collect data and information produced by occurrence reporting, which can be analyzed and processed to improve patient safety.[107] This process presupposes that all members of the organization remain mindful and vigilant, maintaining continuous surveillance over the functioning of the system. Similarly, people who work within the health-care organisation[108] need to believe that they are obliged to report any error. However, the interests of medical institutions should also be taken into account: some errors should be subject to appropriate disciplinary actions.

In the 'just culture' philosophy—as Marx stressed it—individuals are not accountable for system failings when human error is involved, but a 'just culture' does not tolerate reckless behavior, 'gross negligence' or 'wilful misconduct'.[109]

To this end, he identified three categories of behaviour that can affect safety: 'human error', 'at-risk behavior' and 'reckless behavior'.

Any inadvertent action, slip, lapse or mistake falls within the first category (human error). Behavioral choices that increase risk are included in the second category (at-risk behaviour), provided that risk is not recognized or is erroneously believed to be insignificant or justified. Behavioral choices that consciously disregard a substantial, unreasonable and unjustifiable risk fall into the third category (reckless behavior). Grossly negligent conduct, reckless conduct, intentional willful violations are among the latter category.

Striking a balance between the two extremes of punishment and blamelessness, *i.e.* a middle way between punitive and blame free culture, is a difficult task. Nevertheless, the goal of developing a 'just culture' is to move away from an excessively punitive culture.

Subsequently, main studies on 'just culture' were conducted by Professor Sidney Dekker,[110] leading expert on medical accidents and complex system failures. In his well-known book, called "Just Culture. Balancing Safety and Accountability",[111] in

---

[104]Carrol (2010), p. 101 ff.; van Asselt and Renn (2011), p. 431 ff.; Ingratoci (2014), p. 293 ff.

[105]Arghami et al. (2014), p. 143 ff.

[106]Dekker (2003a), 233 ff.; Dusi (2017), p. 81.

[107]Gulseth (2007), p. 17; Van Beuzekom et al. (2010), p. 52 ff.; Van Beuzekom (2012), 12:10.

[108]See Page (2007), and Thew (2013).

[109]Dekker (2003b), p. 83 ff. and Dekker (2005).

[110]Sidney Dekker is a Professor at Griffith University in Brisbane, Australia, where he founded the Safety Science Innovation Lab and at Lund University School of Aviation (Ljungbyhed, Sweden).

[111]Dekker (2016a), chapter 5 and Dekker (2007). See also Boysen (2013), p. 402 ff. and Boysen (2014–2015), issue 7.

which many practical cases from both medical and aviation sectors have been reported, he stresses that the goal is to create, within all complex organisational systems,[112] an environment[113] in which both learning and accountability are well balanced.[114]

In this respect, Dekker notes that a 'just culture' is meant to balance learning from incidents with accountability for their consequences. His advice to creating the basis for a 'just culture' is that firstly incidents need to be normalised and legitimised. Incidents should be seen not as failures or shameful, but as a learning opportunity to improve an organization.

A 'just culture' is a learning culture[115] that is a constantly improving condition, oriented towards patient safety.[116]

The framework of a 'just culture'[117] ensures balanced accountability for both individuals and organizations.

In other words, in a 'just culture' environment, workers are confident that while they will be held accountable for their actions, they will be treated confidentially. However, they also know very well that those who act recklessly, deliberately taking unjustifiable risks, will be punished.[118]

According to Dekker, a 'just culture', in which employees are not afraid to report errors, is a highly successful way to manage safety and crucial for the creation of a 'safety culture'.

Therefore, there is a direct link between a 'just culture' and a 'safety culture'.[119] He thinks that a 'safety culture' cannot flourish without occurrence reporting and information sharing.

'Safety culture', in turn, has been defined as "shared values (what is important) and beliefs (how things work) that interact with an organization's structures and control systems to produce behavioural norms (the way we do things around here)".[120]

Another similar and more sophisticated definition of 'safety culture' is the following, elaborated by Advisory Committee on the Safety of Nuclear Installations (ACSNI):

---

[112]See Almond and Colover (2010), p. 323 ff.

[113]Flin et al. (2000), p. 77 ff., Yule (2003), p. 1 ff., Mearns et al. (2003), p. 641 ff.

[114]Van het Kaar (2010), p. 64 ff.

[115]LeSage et al. (2011), p. 8 ff.

[116]Wise et al. (1993); Cook and Render (2000), p. 791 ff., Sharpe (2003), S1 ff., Youngberg (2010), p. 267 and Youngberg (2012), p. 171; Barnsteiner and Disch (2012b), p. 407 ff.; Dekker (2016b); Woodward (2017).

[117]Sharpe (2004); Katri et al. (2009), p. 313 ff.

[118]Dekker (2008).

[119]See Reinhart (1996), Hudson (1996), Cox and Flin (1998), p. 189 ff.; Wiegmann et al. (2004), p. 117 ff.

[120]Uttal (1983).

the product of individual and group values, attitudes, perceptions, competencies, and patterns of behavior that determine the commitment to, and the style and proficiency of, an organization's health and safety management. Organisations with a positive safety culture are characterised by communications founded on mutual trust, by shared perceptions of the importance of safety and by confidence in the efficacy of preventive measures.[121]

At this point of reasoning, we can note that, beyond its domain of application, a 'just culture' is a culture that recognises and tolerates human fallibility; a culture that moves away from an overly punitive culture and strikes a middle ground between a punitive and blame free culture; a culture of shared accountability, which promotes professionalism; a culture where there is justice; a culture that is keen to learn.

In particular, three key words have emerged so clearly: accountability, learning, and trust. It means that the concept of 'just culture' is characterized by these three fundamental elements.

1.3.2   The Brief Relevance of a 'Just Culture' in the Field of Medical Responsibility in the Italian Legal System: From the 'Balduzzi' Decree to the 'Gelli' Law

Considering the application of 'just culture' principles in the medical sector, it is no coincidence that the Italian recent legislation on medical liability has recently embodied the concept of ordinary fault or negligence in order to exclude health-care professional liability.

For many years, the interpretation provided by case-law played a key role in the field of medical liability.

Until the 1990s, the Italian Court of Cassation had been benevolent towards healthcare professionals, considering that the medical profession is particularly dangerous. Several judgments[122] held that the legal test used to determine the doctor's professional liability would be performed with flexibility, considering the specific characteristics of the medical profession and the difficult situations that could occur.

These particularities required a flexible interpretation of the concepts of negligence, imprudence and malpractice,[123] under Article 43 of the Italian Criminal Code. In this context, the Court considered the degrees of fault referred to in Article 2236 of the Italian Civil Code applicable to the medical liability, both in civil and in criminal proceedings.

---

[121] Advisory Committee on the Safety of Nuclear Installations (ACSNI), Human Factors Study Group: Third report - Organising for safety, HSE Books, 1993.

[122] See Court of Cassation, section IV, criminal division, 6 March 1967 and 21 October 1970, Rivista italiana diritto e procedura penale, 1973, p. 255; 4 February 1972; 17 February 1981; 2 June 1987; No. 10289 of 23 July 1990. They considered Article 2236 and the degrees of fault applicable to the criminal system.

[123] See Nuvolone (1969b); Sharpe (1999), p. 5; Miller and Jones (2011); Amato (2018a), p. 28 ff.

What is the content of these Articles?

Article 43 of this Criminal Code,[124] *inter alia*, states that a crime[125] is intentional when the harmful or dangerous result of an act or omission is foreseen and wanted by the actor as a consequence of his behaviour or failure to act (*dolus directus*).[126] It is negligent when the negative event, even though foreseen, is not desired by the actor, but it occurs because of carelessness, imprudence, lack of skill (generic fault), or failure to observe laws, regulations, orders or instructions (specific fault).[127]

Article 2236 of the Italian Civil Code states the following: when the nature and content of the required services are very technical and involve the solution of very difficult matters, the practitioner does not incur civil liability, except in case of 'willful misconduct' or 'gross negligence'.[128] According to this Article, a person is not liable in case of mere negligence, under these conditions.

It is worth observing that in the ruling No. 166/1973,[129] the Supreme Court applied Article 2236 to medical liability,[130] thus recognising the impact of the private-law criterion of the degree of fault on medical liability. In this decision, however, the Court pointed out that in light of the reasonableness principle this criterion is crucial only in case of 'malpractice', namely where "the indulgence of the court is in direct proportion to difficulties of the task", being instead inapplicable in case of negligence or imprudence.

In this respect, some authors[131] argued that in case of medical 'malpractice', healthcare professionals are liable only for 'gross negligence' (and not for 'ordinary negligence') so as to protect their technical professional discretion, while the general criminal-law rules need be applied in the other two cases of negligence and imprudence. The reason for this different treatment was that only in the first case ('malpractice') could the professional be faced with very difficult problems, not completely explored, debated and solved by science and medical experience.

---

[124]In fact, Article 43 (Mental element of the offense) of the Italian Penal Code says that a crime is negligent when the event, even though foreseen, is not desired by the actor and occurs because of carelessness, imprudence or lack of skill or failure to observe laws, regulations, orders or instructions. See Palombi (2018a), p. 352 ff.

[125]Thomas (2007), p. 57 ff.

[126]In addition, according to Italian doctrine and jurisprudence, there is another type of willful misconduct, called *dolus eventualis* (wilful and wanton conduct or oblique/indirect intent) that is when the wrongdoer foresees the risk and despite this, he acts, accepting the risk of it occurring (see Supreme Court of Cassation, section I, No. 18220 of 11 March 2015 and Section IV, No. 14663 of 30 March 2018). It is called "quasi intent". See Delitala (1932), Ballinari (1992), Prosdocimi (1993), Brunelli (2014), Carboni (2015), chapter 1, para. 1.4; Piqué (2017), p. 1083 ff. See also Oberdiek (1972), p. 389 ff.; Damar (2011), p. 272; Badar (2013), p. 933 ff.; Baker (2016), 121 ff.; Guilfoyle (2016), p. 189.

[127]See Spitzmiller (2011), p. 11.

[128]See Ziccardi (2011), p. 110.

[129]Italian Constitutional Court No.166 of 28 November 1973.

[130]Koch (2011), p. 305; Agich (2012), p. 262; Ferrara et al. (2013), p. 355; Hondus (2014), p. 55.

[131]Crespi (1960), p. 484 ff., Crespi (1973), I, p. 255 ss.; Cattaneo (1958), p. 73; De Simone (1972), p. 825 ff.

Subsequently, since the 1990s and until around the early 2000s, the Supreme Court changed its approach, adopting a stricter position that did not apply the private-law criterion in all cases of criminal responsibility of the doctor or medical staff.

In fact, the judgment No. 4028/1991[132] by the Italian Court of Cassation stated that medical responsibility, in any case, aligns with the criteria laid down in Article 43 of the Italian Criminal Code and not with those contained in Article 2236[133] of the Italian Civil Code.

Similarly, in another ruling of the same Court (No 5278/1995)[134] the following principles were stressed: when professional behaviour affects values acknowledged by constitutional law, such as human life and health, the assessment parameters in order to determine liability could be derived from the rules of criminal law and not from civil law provisions. Therefore, according to this judgment, Article 2236 of the Italian Civil Code, laying down the different degrees of guilt[135] cannot be directly and automatically applied to the field of medical criminal liability. However, when the specific case involves the solution to any special problems or difficulties, the private-law criterion for graduations of guilt severity, contained in Article 2236, could be taken into account by the judge only as an 'experience rule' to determine medical responsibility. In criminal proceedings, the experience rules are to be used to prove elements that, by their special nature, are really hard to prove.[136]

On the contrary, according to Italian legal doctrine,[137] Article 2236 should be interpreted as follows: when professionals are called upon to perform extremely difficult tasks, a level of care proportionate to this difficulty must be adopted. In this case, merely negligent behaviours, referred to in Article 2236, should thus be tolerated in the light of special difficulties of performance.

The arguments brought about by this doctrine do not seem to be convincing. A civil-law provision, such as Article 2236, cannot be applied in criminal matters by means of an analogy mechanism, given the special nature of these rules.[138] In addition, in Italian national law, the graduations of fault test can only be applied in

---

[132]Court of Cassation, criminal division, section IV, No. 4028 of 12 April 1991 and 22 February 1991.

[133]Pontonio and Pontonio (2006), p. 31 ff.

[134]Court of Cassation, section IV, criminal division, No. 5278 of 10 May 1995, Responsabilità civile e previdenza, 1995, p. 903 ff.

[135]Grasso (1979), p. 80, Muscolo (1970), p. 457 ff.

[136]More recently, the Italian Court of Cassation reiterated the general principle that the application of Article 2236 of the Italian Civil Code is limited to compensation for damages in civil proceedings (Court of Cassation, section IV, criminal division, 17 December 2008, No. 46412) and it is not directly applicable in criminal proceedings. In this respect, see also Court of Cassation, section IV, criminal division, No. 11733 of 2 June 1987.

[137]Cattaneo (1982), p. 9; Cian and Trabucchi (1981).

[138]Article 14 of the Italian Civil Code (so-called 'preleggi') does not allow the use of analogy mechanism to fill the regulatory gaps as regards criminal law or in case of exceptional rules. See Morselli (1990), p. 505 ff.; Gallo (2001), p. 1 ff.; Falcinelli (2004), p. 147 ff.

order to determine the amount of the sanction (Article 133 Italian Criminal Code)[139] or to assess the gravity of the infringement, but not to attribute or exclude criminal liability.

The situation changed following the well-known ruling No. 30328 of July 2002 (the so-called 'Franzese Judgment')[140] of the Supreme Court, which represented a milestone for the subsequent jurisprudence.

In this ruling the Court stated that, if the doctor had taken the measures which he negligently omitted to take (counterfactual hypothesis)[141] a positive exit of the disease would have been reached with a high degree of logical probability or rational credibility and the omitted treatment, if it had been performed, would have increased the patient chances of survival.

According to this ruling, the appropriate test to establish the legal liability of doctors has been based on a "counterfactual test",[142] which is based, in turn, on the 'risk increase theory',[143] shared by the majority of Italian judges.

The 'counterfactual test' relied on the '*conditio sine qua non*' theory,[144] known in the Anglo-Saxon legal system as 'but for test'.[145] In civil-law legal traditions, causation reasoning is often used to expand or to stem liability for particular losses flowing from an act of negligence.[146]

Responsibility for a harmful event is attributable to people who have put in place any casualty condition without which the event would not occur. Such verification shall be made *ex post*, by means of the so-called 'counterfactual test'.

According to the "counterfactual test", a conduct or omission can be considered a *conditio sine qua non* of a harmful event when, by removing it from the causal process, the final event does not occur.

In particular, within the field of medical responsibility, the conviction for a negligent crime is founded on the formula 'serious and meaningful (noteworthy) probabilities' of saving the life of the patient, 'beyond a reasonable doubt' (near certainty).[147]

---

[139] At this end, the 'degree of fault' must be interpreted with reference to the degree of predictability of the damaging event (conscious or unconscious fault) and to the expected behaviour.

[140] Judgement of the United Sections of the Court of Cassation, No. 30328 of 11 July 2002, concerning a case of omissive medical responsibility. For a commentary, see Angioni (2006), p. 1282 ff.; Masera (2007), p. 166; Viganò (2006), p. 962, Viganò (2009), p. 1679 ff. and Viganò (2013); Moccheggiani (2017), p. 571 ff.; Mura (2017), p. 3396 ff.; Venturato (2017), p. 1596 ff.; Palombi (2018b), p. 32 ff.

[141] For this concept, see Nocco (2006), p. 1239 ff.; Grassi (2007), p. 404 ff.; Di Landro (2008), p. 181 ff.; Brusco (2008), p. 1875 ff.; Iacoviello (2014), p. 241 ff.

[142] Mandel (2003), p. 245 ff.

[143] Pizzi (2015), p. 1.

[144] Firstly elaborated by Maximilian Von Buri, a German jurist. See Von Buri (1873), and Infantino and Zervogianni (2017), p. 393.

[145] Strachan (1970), p. 386 ff.

[146] Mason (2017), p. 147 ff.; Frecklelton and Mendelson (2017).

[147] Marandola (2012), p. 367 ff.; Polidori (2014), p. 574 ff.; Somma (2014), p. 366 ff., Canzio (2015), p. 193 ff.; Mauceri (2015), p. 114; Paliero (2015), p. 1507 ff.; Saponaro (2018), p. 469 ff.

The elaboration of the 'beyond any reasonable doubt' principle (summarized by the acronym BARD) was intended to make health-care personnel legally responsible for their professional activity.

Because of these divergent judicial interpretations, the Italian lawmakers more recently intervened with the Decree Law No. 158 of 13 September 2012[148] (better known as 'Balduzzi' Decree) on medical liability.[149]

It can be said that this legal instrument represented the first application of 'just culture' principles in the Italian legislation on medical responsibility.

According to Article 3 of this Decree, "Healthcare professionals who, in exercising their profession, comply with the guidelines and follow the best practices recognized by the scientific community,[150] shall not be prosecuted for mere negligence".[151]

Therefore, in order to be discharged of criminal liability, healthcare professionals had to prove that they acted in accordance with the guidelines commonly accepted in the scientific community (authoritative guidelines).[152] This compliance[153] made the action excusable and tolerable *per se*.[154]

The reference to the common concept of 'just culture' is evident. In fact, healthcare professionals who comply with guidelines and best practices are comparable to aviation operators who comply with their training[155] and experience.

To this end, attention has to be drawn to the ruling No. 46753/2013[156] of the Italian Court of Cassation as follows:

> the guidelines under Article 3 of the Balduzzi Law, to be relevant to establish the medical malpractice,[157] should fix the diagnostic and therapeutic standard requirements, consistent with the rules of the best medical science, as a guarantee of the patient's health and they should not be inspired by exclusive logic of managerial economy, as regards to expenditure restraint, in conflict with the patient's care.

---

[148]Laying down urgent provisions to promote the development in Italy of an higher level of health protection", published in G.U. No. 214 of 13 September 2012, converted in the Law No. 189 of 8 November 2012. For a commentary, see Iadecola (2013), p. 549 ff.; Cupelli (2017b), p. 196 ff.; Alagna (2017), p. 1466 ff.; Di Bitonto (2017), p. 3799 ff.

[149]For an overview of the previous legal system, see Zana (1987), p. 420; Ferrari (2004), p. 1492; Marseglia and Viola (2007).

[150]Franzoni (2016), p. 801 ff.; Hall and Roussel (2017), p. 178 ff.

[151]Cupelli (2017c), p. 135.

[152]See Skegg (1998), p. 220 ff.; Pandit (2009), p. 379 ff.

[153]Hudson et al. (2002), p. 1 ff.

[154]Problems were raised in applying this Decree, due to the fact that, in relation to the guidelines, the text referred to a Scientific and Technical Committee for the Guidelines National System (SNLG), established by the so-called 'Sirchia Decree' (Ministerial Decree, Ministry of Health, 30 June 2004), that had never been established. Then the reference to this Committee has been eliminated by the 'Lorenzini draft law' of 2017 on reorganizing the professional associations, which became Law No. 3 of 11 January 2018.

[155]Helreich and Foushee (1993), p. 3 ff.; LeSage et al. (2011), p. 8 ff.

[156]Italian Court of Cassation, section IV, criminal division, No. 46753 of 22 November 2013.

[157]Dauer (2004), p. 185 ff.

In other words, the guidelines must be inspired by the protection needs of the patient and not by different needs.

In another judgment (4468/2015)[158]—under the 'Balduzzi' Decree—the same Court argues that: "the guidelines for therapeutic practices are codified scientific and technological knowledge, able to easily guide treatment decision, to standardize assessments and to minimize subjective decisions of the physician".

Article 3 of the 'Balduzzi' Decree, while excluding the criminal liability for 'simple fault', did not exclude civil liability[159] under Article 2043[160] of the Italian Civil Code (non-contractual liability). In this case, a person who has suffered damages must provide proof that such damages—in accordance to the 'causality credo'[161]—were caused by 'willful misconduct', 'gross negligence' or 'mere negligence' of professionals.

In this respect, Article 3 was regarded as a deviation from previous settled case-law of the Italian Court of Cassation,[162] which clearly excluded the application in criminal matters of the degrees of negligence (gross or ordinary), that typically characterize civil law issues.

In fact, the 'Balduzzi' Decree distinguished between 'mere or ordinary negligence', 'gross negligence' and 'willful misconduct', while did not distinguish between negligence, imprudence and malpractice.

On close examination, 'simple negligence' and 'gross negligence' differ as to the degree of inattention. In fact, in law dictionaries, 'ordinary negligence' is defined as a mere "failure to behave with the level of care that someone of ordinary prudence would have exercised under the same circumstances".[163]

To this end, the particular context and environment in which the event occurred must be taken into account. Therefore, in respect to aviation accidents and incidents, negligence must be measured according to the standards of care (diligence) applicable to a reasonable air traffic controller[164] or pilot,[165] and not referred to the ordinary reasonable man.[166]

---

[158]Italian Court of Cassation, section IV, criminal division, No. 4468 of 29 October 2015.

[159]For the difference between penal and civil fault, in the Italian legal system, see Gallo (1960), p. 624 ff.; Nuvolone (1969a), p. 696; Castronuovo (2009).

[160]Zana (2016), p. 19 ff.; Barbarisi (2017), p. 217 ff. See also Gallone (2017), p. 380 ff.

[161]Masieri (2016), p. 1440 ff.

[162]Italian Court of Cassation, section IV, No. 50078 of 19 October 2017 (deposited on 31 ottobre 2017). For a commentary see Cupelli (2017e), p. 250 ff. and Amato (2018c), p. 75 ff. See also Court of Cassation, section IV, criminal division, No. 28187 of 7 June 2017. For a commentary see Palombi (2018b), p. 32 ff. See Court of Cassation, United Criminal Sections, 21 December 2017, Diritto penale contemporaneo, 11/2017, https://www.penalecontemporaneo.it/upload/9488-su-aiello.pdf.

[163]So Free Legal Dictionary, (2016), Legal Information Institute (LII), Cornell University Law School, Farlex Inc. See Scarabello (2017), p. 40 f.

[164]Comenale Pinto (2009), p. 373 ff.; Rizzo (2009).

[165]Weiner et al. (1993); Catalisano (2015), p. 27 ff.

[166]Schubert (2013), p. 46 ff.

Instead, 'gross negligence'[167] is described as "a conscious and voluntary disregard of the need to use reasonable care, which is likely to cause foreseeable serious injury or harm to persons or property or both".[168] Therefore, 'gross negligence', unlike 'willful misconduct', refers to an act showing a severe and reckless disregard for life or safety of another person.

This gives rise to the following question: can the aforementioned legal definition of 'gross negligence', understood as 'a conscious and voluntary disregard of the need to use reasonable care', be applied to medical liability?

In this regard, the Italian Court of Cassation (No 22405/2015)[169] clarified that the three following parameters can be used in order to evaluate the behaviour of agents and distinguish between gross and ordinary negligence: (a) the divergence between the conduct really exhibited and the behaviour which was to be expected from the agent; (b) the agent's reproachable behaviour; (c) the consciousness to behave dangerously and to bear the risk and its likely consequences accordingly.

A few years after the entry into force of the 'Balduzzi' Decree, its Article 3, which excluded the doctor's responsibility[170] for 'mere negligence' in the case of his compliance with the generally accepted guidelines and good practices, was repealed.

Significant changes to the above mentioned legislation and steps backwards were made by Italian Law No. 24 of 8 March 2017[171] (better known as 'Gelli-Bianco' Law).[172]

---

[167]In Merriam-Webster's (Dictionary of Law, Springfield, Massachusetts, 1996) gross negligence' is described as "failure to exercise the degree of care expected of a person of ordinary prudence in like circumstances in protecting others from a foreseeable and unreasonable risk of harm in a particular situation".

[168]See the Free Legal Dictionary (2005), Legal Information Institute (LII), Cornell University Law School, Farlex Inc.

[169]Italian Court of Cassation, section IV, criminal division, No. 22405 of 8 May 2015, in CED Cassazione penale, 2015, 263736. See also Court of Cassation, criminal division, criminal division, section III, No. 5460 of 4 December 2013, section IV, No. 18430 of 5 November 2013, and No. 47289 of 9 October 2014.

[170]Crespi (1955); Cattaneo (1977); Finucci (1992), p. 420 ff.; Castelletta (2004).

[171]Law No. 24/2017, laying down provisions on safety of care and assisted persons, as well as on the professional responsibility of healthcare professionals (G.U. No. 64 of 17 March 2017). For a commentary, see Sicurezza delle cure e responsabilità sanitaria. Commentario alla legge 8 marzo 2017, n. 24, Quotidiano sanità edizioni, Roma, 2017, *passim*; Hazanlso (2017), p. 75 ff.; Ponzanelli (2017), p. 356 ff.; Caputo (2017b), p. 724 ff. Caputo (2017a), p. 293 ff.; Pardolesi and Simone (2017), p. 161 ff.; Scognamiglio (2017), p. 740 ff.; Caletti and Mattheudakis (2017), p. 85 ff.; Iadecola (2017), p. 53 ff.; Di Giovine (2017), p. 386 ff.; Palma (2017), p. 523 ff.; Velliscig (2018), p. 267 ff.

[172]Regarding relations between the old and the new law under examination, attention has to be drawn to the ruling No. 3/2017 of the Italian Court of Cassation, that clarifies that the 'Gelli' Law "because of its novelty, shall apply only to the events occurred after the entry into force of the new law. To the events which took place before that date, shall continue to apply the repealed Article 3 (1), of the Law No. 189 of 2012, pursuant to Article 2 of the Criminal Code, which excluded criminal law liability of harmful conduct characterized by simple negligence, in a framework regulated by guidelines and best practices recognized by the scientific community". In another

Article 5 of this Law states that, during medical services (therapeutic, diagnostic, preventive, palliative, rehabilitation and forensic medicine), healthcare professionals shall—without prejudice to the specificities of the case—comply with recommendations contained in the guidelines published on the National Health Service website and elaborated by private and public bodies/institutions and scientific societies of healthcare professions[173] listed in a special register established and governed by decree of the Health Minister.[174]

In the absence of such recommendations, health care professionals shall comply with best practices regarding health care.

It is necessary to clarify the difference between guidelines[175] and best practices,[176] since these concepts can be commonly defined as behavioral rules that influence the conduct of healthcare professional. Guidelines have been commonly defined as 'recommendations intended to optimize patient care that are informed by a systematic review of evidence and an assessment of the benefits and harms of alternative options'.[177]

Best practices for health professionals can be defined as 'the conscientious and judicious use of current best evidence in making decisions about the care of individual patients'.[178] These practices derive from evidence, literature and health

---

judgment (No. 50078 of 21 October 2017, subject to commentary by Palombi (2018b), p. 32 ff., the same Court, with reference to the new Law, pointed out that "the eventual impunity, because of its substantial nature, is also applicable, for acts committed before the entry into force of the Law 24/2017, to cases pending before the Court of Cassation and the relevant issues, under Articles 2(4), of the Criminal Code and 129 of the Code of Criminal Procedure, are lodged 'ex oficio' ex Article 609(2), of the Code of Criminal Procedure, even in the case of appeal considered inadmissible. [...] Doubts of interpretation of the new provision are well-known. What is clear, as expressly provided by Article 590 *sexies* (2), is that has been repealed the criminal regime on the decriminalization of simple negligence provided for by the Balduzzi Law, having been repealed the whole Article 3(1). Therefore, there is no longer a question of the degree of fault, except in specific cases, in which the Balduzzi Law constitutes provision more favourable for crimes committed when it was in force, involving aspects of negligence and imprudence qualified as simple fault (extended enforcement of the more favourable regime). It is also clear that the legislator has aimed to limit innovation only to situation of malpractice, i.e. to the fault that is based on the leges artis, considered as not punishable also in case of gross negligence". See Massaro (2017), p. 1 ff.; Carbone (2017), p. 737 ff.

[173]The innovation brought about by the 'Gelli' Law, compared to the 'Balduzzi' Decree, is that the task and the power of producing clinical guidelines and recommendations have been extended to private and public bodies/institutions/associations and no longer reserved for the scientific societies of healthcare, as we have seen. Cf. Cupelli (2017d), p. 1765 ff.; Cembrani (2017), p. 873 ff.; Di Giovine (2017), p. 386 ff.

[174]This provision is governed by the key principles of transparency.

[175]See Bettiol (2017), p. 237 ff.

[176]Roncali (2017), p. 280 ff.

[177]Definition elaborated by the IOM Institute of Medicine (ed), Clinical practice guidelines we can trust, The National Academies Press, Washington, DC, 2011. See also Hudson and Stephens (2000), 1 ff.; Steel et al. (2014), p. 1251 ff.

[178]So Perleth et al. (2001), p. 237 f.

organizations[179] and include any procedure, activity or behaviour based on quality and safety standards.[180]

It should be noted that Italian law attaches greater importance to the guidelines rather than to the best practices.

Therefore, the 'Gelli' Law reaffirms the role of the guidelines as the basis for comparison, already recognized by the 'Balduzzi' Decree.

The most significant change[181] brought about by the new Law, from the viewpoint of the principles of 'just culture', is represented by Article 6, which enacted Article 590-*sexies*[182] into the Italian Criminal Code. The new Article provides medical liability in case of negligent healthcare services[183] that cause death or personal injury of a patient. Article 6 introduces[184] a new case of criminal liability, *i.e.* culpable liability for death or personal injury in the field of health-care[185] into the Italian criminal law system.

The first paragraph of this new Article states that, if involuntary manslaughter[186] or culpable personal injury have been committed whilst exercising the medical profession, the second paragraph shall apply.

Under the second paragraph, any occurrence caused by 'malpractice' shall not be punishable if the medical action has been taken in compliance with recommendations contained in the healthcare guidelines or, in the absence thereof, in clinical care best practices, provided that these recommendations are adapted to the specificities of the particular case. Otherwise, the action is considered a crime and is subject to penalties provided for in Articles 589 and 590 of the same Criminal Code.[187] The first Article refers to the crime of manslaughter, the second one to the crime of culpable injuries.

---

[179]Driever (2002), p. 591 ff. say that "best practice is not a specific practice per se but rather a level of agreement about research-based knowledge and an integrative process of embedding this knowledge into the organization and delivery of health care".

[180]Best practice require specific indicators to be continuously monitored and may vary depending on the sector to which it refers, as well as on the basis of individual areas. So Perleth et al. (2001), p. 237 f.

[181]Cupelli (2017d), p. 1765 ff.; Garzone and Nocco (2017), p. 885 ff.; Hazan (2017), p. 523 ff.; Pavich (2017), p. 2961 ff.

[182]See Article 6 of the 'Gelli' Law. See Caletti and Mattheudakis (2018), p. 25 ff.

[183]D'Alessandro (2017), p. 573 ff.

[184]See Article 590-*sexies* of the Italian Criminal Code, added by Article 6 of the 'Gelli Law'.

[185]Miglio (2016), p. 1250 ff., Dell'Osso (2016), p. 364 ff.

[186]Almond (2012), in particular, chapter 3.

[187]The punishment for (involuntary/culpable) manslaughter is 6 months-5 years' imprisonment, increased up to three time in the case of death of several people (Article 589 of the Criminal Code). The punishment for culpable injuries is a maximum of 3 months' imprisonment (Article 590 of the Criminal Code). In case of serious injury, the penalty is 1–3 months' imprisonment or, alternatively, a fine, excluding the case of serious injuries punishable more severely. In the latter case, the punishment is 3 months-2 years' imprisonment (Article 590(2), of the Criminal Code). See D'Alessandro (2017), p. 573 ff.

According to Italian law, in these cases the prosecutor is required to investigate with a view to gathering relevant evidence of the alleged offence and finding the responsible person. Therefore, within the limits of his/her competence, the prosecutor shall investigate the case in order to reveal the existence or non-existence of a breach of law which entails criminal accountability, identification of the guilty persons and, depending upon the results of investigation,[188] he/she shall decide to prosecute (Article 112 of the Italian Constitution).

The prosecutor is called upon to gather evidence in respect to an objective element (conduct-adverse event-causal link) and a subjective element (negligence/willful misconduct of the health-care professional).[189] Usually he relies on appointed experts for the reenactment of the dynamics of the fatal or adverse event (objective element) and then makes an appropriate assessment of culpability (subjective element). To this end he should not look at the result, but rather at the action, or better at the intent behind the action.[190]

The application of 'just culture' principles do not guarantee immunity from criminal investigation for this purpose, but it allows the judicial authority to evaluate very carefully the behavior of the doctor, considering that health-care is an extremely complex and risky activity.

Having said that, under the wording of the new Article 590-*sexies*,[191] impunity of healthcare professional depends on three conditions:

(a) adverse event caused by 'malpractice', with implied exclusion of the other two situations, negligence and imprudence, referred to in Article 43 of the Italian Criminal Code[192];
(b) compliance with the guidelines or, in the absence thereof, best practices;
(c) suitability of recommendations contained in the guidelines for the specificities of the case.

Firstly, the concept of 'malpractice' must be clarified, as referred to in *(a)*. In general, it refers to any misconduct consisting of an unreasonable lack of skill and education in carrying out professional duties and performances.[193] It is a violation of the conduct *leges artis*,[194] intended as unwritten technical rules that the professional is required to observe. In the healthcare sector, 'malpractice' materializes when the

---

[188]On the issue of the prosecutor's discretion, see Cordero (1957), p. 158 ff.; Luparia (2002), p. 1751 ff.; Vicoli (2003), p. 251.

[189]Panelli and Scarabello (2013), p. 18, 59 f.

[190]Brüggen (2013), p. 44 f.

[191]Cupelli (2018), p. 246 ff.

[192]In fact, Article 43 of the Italian Penal Code says that a crime is negligent when the event, even though foreseen, is not desired by the actor and occurs because of carelessness, imprudence or lack of skill (or failure to observe laws, regulations, orders or instructions).

[193]Diehl (1989).

[194]Drobnig et al. (2017), p. 103; Buzzoni (2007), p. 60; Lobato de Faria (2010), p. 99; Nefeli Gribaudi (2012), p. 55; Risicato (2013), p. 15 f., Winger et al. (2018), p. 784; Terrizzi (2018), p. 93 ff.

doctor or other healthcare professional acted outside of the minimum level of medical experience and culture.

However, there is a great difficulty in drawing a clear dividing line between negligence, imprudence and malpractice,[195] which often overlap with each other.

'Malpractice' differs from negligence that is a failure to exercise reasonable care, or attention in performing specific tasks and also from 'imprudence', which is an impulsive behaviour, characterized by acting without any concern for the consequences and with lack of judgment.

Under Article 590-*sexies*, if the conduct is characterized by 'malpractice', the health-care personnel shall not be held liable, while if it is characterized by negligence or imprudence the medical professional shall be held liable.

What is the reason of this different treatment in case of 'malpractice'? The difference depends on the special difficulties of the service involved, as mentioned above.

According to the condition that we have marked with the letter *b), i.e.* compliance with the clinical official guidelines or best practices in order to exclude health-care professional liability, the new Law has limited the crucial role attributed to these guidelines[196] by the 'Balduzzi' Decree, confining it to a single criterion used by the courts for determining medical responsibility.

In addition, considering that the scope of the 'Gelli' Law has been restricted to the 'malpractice' case, Article 590-*sexies* refers exclusively to the guidelines containing rules of medical expertise.[197]

Article 590-*sexies* repeals the provision already contained in Article 3 of the 'Balduzzi' Decree, which excluded the application of criminal sanctions in the case of compliance with medical guidelines, but it also assigns relevance to best practices.[198]

It must be pointed out that guidelines often contain very flexible recommendations. For instance, in the case of treating an orthopedic disease, the guidelines leave the doctor free to choose between conservative and surgical intervention. If the healthcare professional opts for the conservative treatment, after having verified the adequacy of this recommendation to the specific case, and then the patient suffers serious harm as a result of this choice, it seems, *prima facie,* that all conditions mentioned in Article 590-*sexies* are met. In fact, in this case, professionals followed the guidelines and therefore they should not be accused of 'malpractice'.[199]

On close examination, a case of 'malpractice' may occur in the case where the health professional follows the flexible and discretionary recommendation contained in the guidelines, ignoring the fact that the scientific literature suggests surgery in this particular case. Indeed, as clarified by the Italian Court of Cassation in its ruling

---

[195]Caletti and Mattheudakis (2017), p. 101 ff.

[196]Italian Court of Cassation, section IV, criminal division, No. 3 of 20 April 2017.

[197]Bartoli (2018), p. 233 ff.

[198]Piras (2018), p. 4 ff.

[199]See Todeschini (2017).

No. 35992/2012,[200] the guidelines must be complemented and combined with the scientific literature.

Therefore, in this situation, the guidelines were not fully respected.

Finally, according to the letter c), despite compliance with clinical practice guidelines, if the latter were not appropriate and applicable in that particular case, the healthcare personnel shall likewise be held liable. Therefore, this provision requires the physician to verify the guidelines and their application to the specific case.

It is necessary to comment on several points. The conditions mentioned in points b) and c) seem to conflict with that mentioned in point *a)*. In fact, if guidelines are fully observed and they are suited to the specificities of the particular case, this is not a case of 'malpractice'. This rather seems like a contradiction in terms, considering that guidelines represent a well-established parameter for blaming doctors for medical errors and mistakes in health care. Therefore, 'malpractice' should be automatically excluded by compliance with guidelines and by their appropriateness for the particular case. Actually, in the case in which both compliance and adequacy conditions have been met, in our opinion, there is no place for generic fault, under Article 43 of the Italian Criminal Code.[201]

In very blunt terms, the new Law, on the one hand, abolishes the crucial distinction between gross and ordinary (not serious) negligence, failing to distinguish the degrees of fault and, on the other hand, introduces a special preferential regime of exemption of liability only in the case of 'malpractice', excluding the cases of negligence and imprudence. Consequently, impunity of 'malpractice', under the conditions laid down in Article 590-*sexies* of the Criminal Code, is now granted even in the case of 'gross negligence'.

Clearly this Law intends to favour the case of 'malpractice' and to eliminate any reference to different levels of fault.

The recent ruling No. 8770/2017[202] of the United Sections of the Italian Court of Cassation is particularly interesting to understand the true extent of the 'Gelli' Law. It clarifies that the liability of the health-care professional for a patient's death or personal injury, has to be interpreted as meaning that the doctor is held responsible for culpable behaviour in several cases.

In particular, the healthcare professional is responsible for death and personal injury incurred by negligence in the exercise of the profession in the following cases: if the event occurred due to (simple or gross) 'negligence' or 'imprudence'; if the event occurred due to 'malpractice', when the concrete case does not fall under any guidelines or good clinical practices; if the event occurred due to 'malpractice' in identifying and choosing the guidelines and good practices, which are not

---

[200]Italian Court of Cassation, section IV, criminal division, 19 September 2012.

[201]See Cavaliere (2017).

[202]Italian Court of Cassation, Joint Sections, No. 8770 of 21 December 2017, Guida al diritto, 2018, 12, p. 40. For a commentary, see Brusco (2007), p. 205 ff. and Brusco (2018), p. 646 ff.; Amato (2018b), p. 299; Palombi (2018b), p. 32 ff.; Risicato (2018), p. 948 ff.; Palmieri (2018), p. 235 ff.

appropriated to the circumstances of the particular case, or if the event occurred due to 'malpractice' in implementing appropriate recommendations, guidelines or good clinical practices, taking into account of the degree of risk involved and of the special difficulties of medical practice.

According to this judgment,[203] the physician should be prudent about following the further progress of the case before it. He/she must very knowledgeable on the *leges artis*, able to diagnose, be up-to-date on new scientific findings and on their application, able to make *ex ante* choices appropriate to the specific case. If an adverse or fatal event nevertheless occurs as a result of a medical behaviour, the error qualified as 'malpractice' shall not be punished, because it is considered as a very small deviation from the guidelines.[204]

Despite this broad interpretation of the Gelli Law by the Supreme Court there has been a resulting de-evolution of the medical Italian legal system, which has worsened the position of the healthcare professional,[205] by eliminating the decriminalization of 'simple negligence' and, consequently, the relevance of the 'just culture' guiding principles.

On the contrary, in aviation law these principles have been incorporated in many international instruments and European legislation, as will be described below.

References

Agich GJ (ed) (2012) Responsibility in health care. Kluwer, Dordrecht

Alagna R (2017) La controriforma della colpa penale dell'attività medica. Responsabilità civile e previdenza, pp 1466–1489

Almond P (2012) Corporate manslaughter and regulatory reform. Palgrave Macmillan, London

Almond P, Colover S (2010) Mediating punitiveness; understanding public attitudes towards work-related fatality cases. Eur J Criminal 7(5):323–338

Amalberti R (2001) The paradoxes of almost totally safe transportation systems. Safety Sci 37:109 ff

Amato G (2018a) Conclusione giusta in linea con la norma e contro le negligenze. Guida al diritto 12:28–32

Amato G (2018b) Conclusione giusta in linea con la norma e contro le negligenze. Archivio giuridico circolazione assicurazione e resonsabilità 4:299 ff

Amato G (2018c) Sussiste la colpa quando è ravvisato un errore inescusabile. Guida al diritto 1:74–77

Anderson M (2015) Behavioural safety and major accident hazards: magic bullet or short in the dark. In: Conference Proceedings, Hazards XVIII Symposium, 24 November 2004, IChemE/ UMIST, Manchester

Angioni F (2006) Note sull'imputazione dell'evento colposo con particolare riferimento all'attività medica. Studi in onore di Giorgio Marinucci, II, Giuffré, Milano, pp 1279–1334

---

[203]See para. 9.1 of the Court of Cassation judgement No. 8770 of 21 December 2017, cited above.

[204]See also Massaro (2017), p. 12; Cupelli (2017c), p. 135 ff., Cupelli (2017a), p. 244 ff. and Cupelli (2017e), p. 250.

[205]So Tirella (2017).

Arghami S, Nouri Parkestani H, Alimohammadi I (2014) Reliability and validity of a safety climate questionnaire. J Res Health Sci 14(2):140–145

Arnesen SA (ed) (1995) Interact '95. Chapman and Hall, London

Badar ME (2013) The concept of Mens Rea in international criminal law: the case for a unified approach. J Int Crim Just 11(4):933–936

Baker DJ (2016) The right not to be criminalized: demarcating criminal law's authority. Routledge, Taylor & Francis, London, New York

Balcerzak T (2017) A just culture? Conflict of interest in the investigation of aviation accidents. Sci J Silesian Univ Technol Series Transp 94:5–17

Ballinari L (1992) Per dolo eventuale: satira su un giudizio penale e una reclusione perpetua. Svizzera, Lugano

Barbarisi A (2017) L'onere della prova nella responsabilità sanitaria. Contratti, Contratti 2:217–230

Barnsteiner J, Disch J (2012a) A just culture for nurses and nursing students. In: Barnsteiner J, Disch J (eds) Second generation QSEN, an issue of nursing clinics of North America. Elsevier, Philadelphia, pp 407 ff

Barnsteiner J, Disch J (eds) (2012b) Second generation QSEN, an issue of nursing clinics of North America, Elsevier, Philadelphia

Bartoli R (2018) Riforma Gelli-Bianco e Sezioni Unite non placano il tormento: una proposta per limitare la colpa medica. Diritto penale contemporaneo 5:233–248

Benkö M (ed) (2012) Essential air and space law, vol 10. Elven Law, Utrecht

Berlinger N (2005) After harm: medical error and the ethics of forgiveness. Johns Hopkins University, Batilmore

Bettiol L (2017) Riforma Gelli-Bianco: il ruolo delle linee guida nel giudizio di responsabilità penale in campo sanitario. Foro italiano, pp 236–241

Beveridge A (ed) (2009) Forensic investigation of explosions. Taylor & Francis, London

Bijlsma F (2013) Justice and safety, vol 18. Hindlight, pp 62–65

Boysen PG (2013) Just culture: a foundation for balanced accountability and patient safety. Ochsner J 13(3):400–406

Boysen PG (2014–2015) Just culture. OT Safety Bull Focus Enforcement 7:1 ff

Brüggen J (2013) Why we need positive examples in our just culture. HindSight 18:44–45

Brunelli D (2014) Il "mistero" del dolo eventuale: scritti dal dibattito svoltosi a Perugia, il 27 gennaio 2012. Giappichelli, Torino

Brusco C (2007) Cassazione e responsabilità penale del medico. Tipicità e determinatezza nel nuovo art. 590-sexies c.p. Diritto penale contemporaneo 11:205 ff

Brusco C (2008) Applicazioni concrete del criterio della probabilità logica nell'accertamento della causalità. Cassazione penale, pp 1875–1886

Brusco C (2018) Responsabilità medica penale: le Sezioni Unite applicano le regole sulla responsabilità civile del prestatore d'opera. Diritto penale e processo, pp 646–654

Buzzoni A (2007) Medico e paziente. Le responsabilità civili e penali del medico e dell'équipe medica. edizioni Fag, Milano

Byrne G (2002) Flight 427: anatomy of an air disaster. Copernicus Books, New York

Caldwell CL (2018) Safety culture and high-risk environments: a leadership perspective. CRC Press, Taylor & Francis, Boca Raton

Caletti GM, Mattheudakis ML (2017) Una prima lettura della legge Gelli-Bianco nella prospettiva del diritto penale. Diritto penale contemporaneo 2:85–107

Caletti GM, Mattheudakis ML (2018) La fisionomia dell'art. 590-sexies c.p. dopo le Sezioni Unite tra "nuovi" spazi di graduazione dell'imperizia e "antiche" incertezze. Diritto penale contemporaneo 4:25–46

Canzio G (2015) Ragioni, verità e dubbio nel labirinto del processo penale. Giustizia penale, pp 193–198

Caputo M (2017a) La responsabilità penale dell'esercente la professione sanitaria dopo la L. n. 24 del 2017. . .quo vadis? Primi dubbi, prime risposte, secondi dubbi. Danno e responsabilità, pp 293–300

Caputo M (2017b) Promossa con riserva. La legge Gelli-Bianco passa l'esame della Cassazione e viene "rimandata a settembre" per i decreti attuativi. Rivista italiana di medicina legale, pp 724–743

Carbone V (2017) Legge Gelli: inquadramento normativo e profili generali. Corriere giuridico, pp 737–739

Carboni L (2015) Il dolo eventuale dopo la sentenza Thyssenkrupp. Key, Vicalvi (Frosinone)

Carnino A, Weimann G (eds) (1995) Proceedings of the international topical meeting on safety culture in nuclear installations. American Nuclear Society of Austria, Vienna

Carrol R (2010) Risk management handbook for health care organizations, 6th edn. Jossey-Bass, Chicago

Castelletta A (2004) Responsabilité médicale: droit des maladies. Dalloz, Paris

Castronuovo D (2009) La colpa peale. Giuffrè, Milano

Catalisano D (2015) The remote pilot operator in the aviation industry. Pilot perspective. In: Pellegrino F (ed) Legislation and regulation of risk management in aviation activity, vol II. Giuffré, Milano, pp 27–44

Catino M (2013) Organizational Myopia. Problems of rationality and foresight in organizations. Cambridge University Press, Cambridge

Catino M, Patriotta G (2013) Learning from errors: Cognition, emotions and just culture at the Italian Air Force. Organ Stud 34(4):437–467

Cattaneo G (1958) La responsabilità del professionista. Giuffré, Milano

Cattaneo G (1977) La responsabilità del professionista. Giuffré, Milano

Cattaneo G (1982) La responsabilità medica nel diritto italiano. In: AA.VV (ed) La responsabilità medica. Giuffré, Milano, pp 9–23

Cavaliere A (2017) Responsabilità medica alla luce della Riforma. Diritto.it

Cembrani F (2017) Su alcuni snodi critici della legge "Gelli-Bianco". Rivista italiana di medicina legale, pp 873–879

Challinor CAS (2016) ICAO Annex 13, Chicago Convention and the admissibility of air accident reports in litigation: is it time for ICAO to change course? International Institute of Air & Space Law, Montreal

Cian G, Trabucchi A (1981) Commentario breve al codice civile. Cedam, Padova

Comenale Pinto MM (2009) Responsabilità nel controllo del traffico aereo. In: Rizzo MP (ed) La gestione del traffico aereo: profili d diritto internazionale, comunitario e interno. Giuffré, Milano, pp 373–386

Comenale Pinto MM (2015) Le raccomandazioni di sicurezza. In: Pellegrino F (ed) Legislation and regulation of risk management in aviation activity, vol II. Giuffré, Milano, pp 83–96

Contissa G, Sartor G, Lanzi P, Marti P, Tomasello P (2012) Liabilities and automation in aviation. In: Proceedings SESAR Innovation Days, 27th–29th November 2012

Contissa G, Laukyte M, Sartor G, Masutti A, Lanzi P, Marti P, Tomasello P, Schebesta H (2013) Liability and automation: issues and challanges for socio-technical systems. J Airsp Oper 2:79–98

Cook R, Render M (2000) Gaps in the continuity of care and progress on patient safety. Br Med J 320:791–794

Cooper D, Findley L (2013) Strategic safety culture roadmap. BSMS (B-Safe Management Solutions) Inc., Franklin

Cooper MD (2000) Towards a model of safety culture. Safety Sci 36:111–136

Cordero F (1957) Le situazioni soggettive nel processo penale. Giappichelli, Torino

Cox S, Flin R (1998) Safety culture: Philosopher's stone or man of straw? Work Stress 12 (3):189–201

Crespi A (1955) La responsabilità penale nel trattamento medico chirurgico con esito infausto. Priulla, Palermo

Crespi A (1960) Il grado della colpa nella responsabilità professionale del medico chirurgo. Scuola positiva, pp 484 ff

Crespi A (1973) La colpa grave nell'esercizio dell'attività medico chirurgica. Rivista italiana diritto e procedura penale I:255 ff

Cromie S, Bott F (2016) Just culture's "line in the sand" is a shifting one; an empirical investigation of culpability determination. Saf Sci 86:258–272

Cromie S, Liston P, Ross D, Corrigan S, Vani L, Lynch D, Demosthenous S, Leva C, Kay A, Demosthenousb V (2013) Human and organisational factors training as a risk management strategy in an aviation maintenance company. Chem Eng 33:445–450

Cupelli C (2017a) Cronaca di un contrasto annunciato: la legge Gelli-Bianco alle Sezioni Unite. Diritto penale contemporaneo 11:244–286

Cupelli C (2017b) Il perimetro applicativo della Legge Balduzzi: aperture giurisprudenziali "vs." restrizioni normative? Processo penale e giustizia, pp 196–204

Cupelli C (2017c) La legge Gelli-Bianco nell'interpretazione delle Sezioni Unite: torna la gradazione della colpa e si riaffaccia l'art. 2236 c.c. Diritto penale contemporaneo 12:135–138

Cupelli C (2017d) La responsabilità penale degli operatori sanitari e le incerte novità della legge Gelli-Bianco. Cassazione penale, pp 1765–1778

Cupelli C (2017e) Quale (non) punibilità per l'imperizia? La Cassazione torna sull'ambito applicativo della legge Gelli. Diritto penale contemporaneo 11:250–255

Cupelli C (2018) L'art. 590-sexies c.p. nelle motivazioni delle Sezioni Unite: un'interpretazione 'costituzionalmente conforme' dell'imperizia medica (ancora) punibile. Diritto penale contemporaneo 3:246–258

Czech BA, Groff L, Strauch B (2014) Safety Cultures and Accident Investigation: Lessons Learned from a National Transportation Safety Board Forum, 13–16 October 2014, Adelaide, Australia

D'Alessandro F (2017) La responsabilità penale del sanitario alla luce della riforma "Gelli-Bianco". Diritto penale e processo, pp 573 ff

Damar D (2011) Wilful misconduct in international transport law. Springer, Hamburg

Daniels S (2017) Corporate manslaughter in the maritime and aviation industries, Lloy's practical shipping guides. Informa Law from Routdlege, New York

Dauer EA (2004) Ethical misfits: mediation and medical malpractice litigation. In: Sharpe VA (ed) Accountability: patient safety and policy reform. Georgetown University Press, Washington, pp 185–201

Davina A, Braithwaite J, Sandall J (eds) (2016) The sociology of healthcare safety and quality. Wiley Blackwell, Oxford, pp 31 ff

De Simone V (1972) Sulla colpa professionale. Giustizia penale, pp 825 ff

Dekker S (2003a) Failure to adapt or adaptations that fail: contrasting models on procedures and safety. Appl Ergon 34(3):233–238

Dekker S (2003b) When human error becomes a crime. Human Factors Aerosp Saf 3(1):83–92

Dekker S (2005) The questions about human error, a new view of human factors and system safety. Taylor and Francis, London

Dekker S (2007) Just culture. Balancing safety and accountability, 1st edn. Ashgate, Aldershot (Hampshire)

Dekker S (2008) Just culture: who gets to draw the line? Cognition technology work. Springer, London

Dekker S (2016a) Just culture: restoring trust and in your accountability organization, 2nd edn. CRC Press, Taylor & Francis, Abingdon-on-Thames

Dekker S (2016b) Patient safety: a human factors approach. CRC Press, Boca Raton

Delitala G (1932) Dolo eventuale e colpa cosciente. Annuario dell'Università Cattolica del Sacro Cuore, Milano

Dell'Osso AM (2016) In tema di colpa medica. Rivista italiana di medicina legale e del diritto in campo sanitario, pp 362–367

Di Bitonto ML (2017) Professione medica e procedimento penale: le novità dopo la legge n. 24/2017. Cassazione penale 10:3799–3808

Di Giovine O (2017) Colpa penale, "legge Balduzzi" e "disegno di legge Gelli-Bianco": il matrimonio impossibile tra diritto penale e gestione del rischio clinico. Cassazione penale Cassazione penale 1:386–404

Di Landro A (2008) Interruzione del nesso causale e accertamento della causalità "modello Franzese". Foro italiano, pp 181–184

Diehl A (1989) Human performance/system safety issues in aircraft accident investigation and prevention. In: Proceedings of 11th International Symposium on Aviation Psychology, Columbus, Ohio

Dismukes RK, Berman BA, Loukopoulos LD (2017) The limits of expertise. Rethinking pilot error and the causes of airline accidents. Ashgate, Farnham

Douglas M (1992) Risk and blame: essays in cultural theory. Routledge, London

Driever MJ (2002) Are evidenced-based practice and best practice the same? West J Nurs Res 24 (5):591–597

Drobnig U, David R, Egawa HH, Graveson R, Knapp V, Von Mehren AT, Noda Y, Rozmaryn S, Tschchikvadze VM, Valladão H, Yntema H (eds) (2017) International Encyclopedia of comparative law. International association of legal sciences, vol 4. Martinus Nijhoff, Brill, Leiden

Dusi G (2017) Just Culture Safety Report: atteggiamenti e comportamenti dei professionist. Youcanprint Self-Publishing, Tricase

Falcinelli D (2004) Analogia: il limite logico del sistema penale. Giurisprudenza italiana, pp 147 ff

Ferguson J, Fakelmann R (2005) The culture factor. Front Health Serv Manage 22:33 ff

Ferrara SD, Boscolo-Berto R, Viel G (eds) (2013) Malpractice and medical liability: European State of the art and guidelines. Springer, Berlin

Ferrari S (2004) Sulla valutazione della responsabilità medica per colpa. Giurisprudenza italiana 7:1492–1494

Festinger LA (1957) Theory of cognitive dissonance. Stanford University Press, Stanford

Finucci G (1992) Riflessioni sulla responsabilità professionale del medico nella complessa situazione sanitaria moderna. Nuovo diritto, pp 420 ff

Flin R, Mearns K, O'Conner P, Bryden R (2000) Measuring safety climate: identifying the common features. Saf Sci 34:177–192

Flin R, O'Connor P, Crichton M (2008) Safety and sharp end: a guide to non-technical skills. Ashgate, Farnham

Ford C, Jack T, Crisp V, Sandusky R (1999) Aviation accident casual analysis. Advances. Aviation Safety Conference Proceedings, pp 343 ff

Fragali M (1967) Regolamento giuridico dell'aviazione e allegati tecnici di Chicago. Diritto aeronautico, pp 159–162

Franzoni M (2016) Colpa e linee guida. Danno e responsabilità 8/9:801–806

Frecklelton I, Mendelson D (eds) (2017) Causation in law and medicine. Taaylor & Francis, Ashgate Publishing, Aldershot

Gallo F (2001) Norme penali e norme eccezionali nell'art. 14 delle "disposizioni sulla legge in generale". Rivista di diritto civile 1:1–28

Gallo M (1960) Colpa penale (diritto vigente). Enciclopedia del diritto. Giuffrè, Milano, pp 624–643

Gallone G (2017) Legge Gelli: il medico "strutturato" risponde (ma, in realtà, già rispondeva) ex art. 2043 c.c. e non già ex art. 1218 c.c. Archivio giuridico della circolazione e dei sinistri stradali 5:379–387

Garzone FP, Nocco BA (2017) La responsabilità penale sanitaria nel passaggio dalla Legge Balduzzi alla Legge Gelli: "se vogliamo che tutto rimanga com'è, bisogna che tutto cambi". Rivista penale 10:885–888

Giannini A (1944) I trasporti aerei internazionali della convenzione di Chicago 1944. Rivista aeronautica, pp 101–103

Giannini A (1946) La convenzione di Chicago 1944 sull'aviazione civile internazionale. Rivista di diritto comm.erciale I, pp 83–92

Giannini A (1949) L'ammissione dell'Italia alla convenzione di Chicago (1944) sull'aviazione civile internazionale. Rivista di diritto della navigazione, I, pp 41–48

Giannini A (1952) Gli emendamenti della convenzione di Chicago (1944) relativa all'aviazione civile internazionale. Rivista aeronautica, pp 729–731

Giannini A (1953a) Il regolamento della navigazione aerea nella convenzione di Chicago 1944. Rivista aeronautica, pp 11–16

Giannini A (1953b) La convenzione di Chicago e la sua tecnica. Rivista aeronautica, pp 269–271

Giannini A (1953c) La convenzione di Chicago 1944 sull'aviazione civile internazionale. Roma

Glendon AI, Clarke SG, McKenna EF (2006) Human safety and risk management. CRC Press, Taylor & Francis, Boca Raton

Grassi M (2007) Nesso di causalità nella responsabilità medica: note a margine di una sentenza milanese. Il Foro ambrosiano, pp 404–414

Grasso G (1979) La responsabilità penale nell'attività medico-chirurgica: orientamenti giurisprudenziali sul "grado" della colpa. Rivista italiana di medicina legale 1:80–91

Green J (2003) The ultimate challenge for risk technologies: controlling the accidental. In: Summerton J, Benner B (eds) Constructing risk and safety in technological practice. Routledge, London, pp 29–42

Guilfoyle D (2016) International criminal law. Oxford University Press, Oxford

Guldenmund FW (2000) The nature of safety culture: a review of theory and research. Saf Sci 34:215–257

Gulseth M (2007) Managing anticoagulation patients in the hospital: the inpatient anticoagulation service. American Society of Health-System Pharmacists, Bethesda, pp 17 ff

Haanappel PPC (2003) The law and policy of air space and outer space: a comparative approach. Kluwer Law International, The Hague

Hale A, Wilpert B, Freitag M (1997) After the event: from accident to organisational learning. Pergamon Press, New York

Hall HR, Roussel LA (eds) (2017) An integrative approach to research. Administration and Practice, Jones and Bartlett Learning, Burlington

Harris D (2011) Human performance on the flight deck. Ashgate, Furnham

Harris D, Li W-C (2006) Where safety culture meets national culture: the how and why of the China Airlines CI-611 accident. Human Factors Aerosp Saf 5(4):345–353

Haslam RA, Hide SA, Gibb AGF, Gyi DE, Pavitt T, Atlinson S, Duff AR (2004) Contributing factors in construction accidents. Appl Ergon 36:401–415

Havel BF, Sanchez GS (2014) The Principles and practice of international aviation law. Cambridge University Press, Cambridge

Hazan M (2017) Alla vigilia di un cambiamento profondo: la riforma della responsabilità medica e della sua assicurazione. Danno e responsabilità, pp 75–91

Hazanlso M (2017) Alla vigilia di un cambiamento profondo: la riforma della responsabilità medica e della sua assicurazione (DDL Gelli). Danno e responsabilità, pp 75 ff

Heinrich HW, Petersen D, Roos N (1980) Industrial accident prevention: a safety management approach, 5th edn. McGraw-Hill, New York

Helreich RL, Foushee HC (1993) Why crew resource management? Empirical and theoretical bases of human factors training in aviation. In: Weiner EL, Kanki BG, Helmreich RL (eds) Cockpit resource management. Academic Press, San Diego, p 3 ff

Hobbs A, Robertson MM (1995) Human factor in airline maintenance. In: Workshop Report, Proceedings of the Third Australian Aviation Psychology Symposium, pp 468 ff

Hollnagel E (2004) Barriers and accident prevention, Ashgate, Aldershot

Hollnagel E (2009) The ETTO principle: efficiency-thoroughness trade-off. Ashgate, Farnham

Hollnagel E (2013) Is justice really important for safety? HindSight 18:10–13

Hondus E (2014) The development of medical liability. Cambridge University Press

Hudson PTW (1996) Psychology and safety. Rijks Universiteit, Leiden

Hudson PTW (2001) Safety Culture: Theory and Practice. In: The Human Factor in System Reliability – Is Human Performance Predictable? (Les Facteurs humains et la fiabilité des systèmes – Les performances humaines, son telles prévisibles?). Papers presented at the Human Factors and Medicine Panel (HFM) Workshop held in Siena, Italy, from 1-2 December 1999, published in NATO series RTO (Research and Technology Organization) MP-032, p 8

Hudson PTW (2003a) Applying the lessons of high-risk industries to medicine. Qual Saf Health Care 12:7–12

Hudson PTW (2003b) Aviation safety culture. J Aviat Manag 4:27–48

Hudson PTW, Stephens D (2000) Cost and benefit in HSE: a model for calculation of cost-benefit using incident potential. In: Proceedings 5th SPE International Conference on Health, Safety and Environment in Oil and Gas Production and Exploration. Leiden, SPE, Richardson, Texas

Hudson PTW, Parker D, Lawton R, van der Graaf GC (2002) Managing non-compliance: moving from theory to practice. In: Proceedings 6th SPE International Conference on Health Safety and Environment in Oil and Gas Exploration and Production, Society of Petroleum Engineers, Richardson, Texas

Iacoviello FM (2014) La "Franzese": ovvero quando buone teorie producono cattiva giustizia. Critica del diritto 3:241–258

Iadecola G (2013) Brevi note in tema di colpa medica dopo la c.d. legge Balduzzi. Rivista italiana di medicina legale 1:549–553

Iadecola G (2017) Qualche riflessione sulla nuova disciplina della colpa medica per imperizia nella legge 8 marzo 2017 n. 24 (legge cd. Gelli-Bianco). Diritto penale contemporaneo 6:53–66

Infantino M, Zervogianni E (eds) (2017) Causation in European tort law. Cambridge University Press, Cambridge, pp 393 ff

Ingratoci C (2014) Notification and reporting of aircraft accidents or incidents. In: Pellegrino F (ed) Legislation and regulation of risk management in aviation activity, vol I. Giuffré, Milano, pp 293–317

Jeffries FL (2011) Predicting safety related attitudes in the workplace: the influence of moral maturity and emotional intelligence. Institute of Behavioral and Applied Management, Anchorage (Alaska)

Johnson C (2005) Visualizing the relationship between human error and organizational failure. Department of Computing Science, University of Glasgow, Glasgow. http://www.dcs.gla.ac.uk/

Johnson CW (1995) Decision theory and safety-critical interfaces. In: Nordby K, Helmersen PH, Gilmore D, Johnston N (eds) Do blame and punishment have a role in organisational risk management? vol 15. Flight Deck, p 33 ff

Johnston N, McDonald N, Fuller R (eds) (1994) Applications of psychology to the aviation system. Averbury Aviation, Aldershot

Katri N, Brown GD, Hicks LL (2009) From a blame culture to a just culture in health care. Health Care Manag Rev 34(4):312–322

Kennedy R, Kirwan B (1995) The failure mechanisms of safety culture. In: Carnino A, Weimann G (eds) Proceedings of the international topical meeting on safety culture in nuclear installations, American Nuclear Society of Austria, Vienna, 24–28 April 1995, 27(13):281–290

Khatib TM (1996) Organizational culture, subcultures, and organizational commitment. Iowa State University, Ames. https://lib.dr.iastate.edu/rtd

Kim K (1989) Human reliability model with probabilistic learning in continuous time domain. Microelectron Reliab 29(5):301–311

Koch BA (ed) (2011) Medical liability in Europe: a comparison of selected jurisdictions. De Gruyter, Vienna

Kohn LT, Corrigan JM, Donaldson MS (eds) (2000) To Err is human: building a safer health system. Institute of Medicine (US), Committee on Quality of Health Care in America, National Academies Press, Washington DC, pp 1–8

Lawrenson AJ, Braithwaite GR (2018) Regulation or criminalisation: what determines legal standards of safety culture in commercial aviation. Safety Sci 102:251–262

Leape L (1994) Error in medicine. J Am Med Assoc 272(23):1851–1857

Leplat J (1982) Accidents and incidents production: methods of analysis. J Occup Accid 4 (2–4):299–310

LeSage P, Dyar JT, Evans B (2011) Crew resource management: principles and practice, developing a culture for open communication. Jones and Bartlett Publishers, Sudbury

Li WC, Harris D (2005) HFACS analysis of ROC air force aviation: reliability analysis and cross-cultural comparison. Int J Appl Aviat Stud 5:65–81

Lintner T, Dunlap T (2013) Just culture and American jurisprudence. Hinsight 18:52–55

Liu H, Zhang C (2007) Study on safety comprehensive assessment of construction site in man-machine-environment system. J Chongqing Jianzhu Univ, pp 107–111

Lobato de Faria P (2010) Medical law in Portugal. Kluwer, Alphen aan den Rijn, pp 99 ff

Long S, Dhillon BS (eds) (2016) Man-machine-environment system engineering. In: Proceedings of the 17th International Conference on MMESE, Springer, Singapore

Lu AC-J (2013) International airline alliances: EC competition law/US. Kluwer Law International, The Hague

Luparia L (2002) Obbligatorietà e discrezionalità dell'azione penale nel quadro comparativo europeo. Giurisprudenza italiana 8:1751–1758

Mandel DR (2003) Effect of counterfactual and factual thinking on causal judgements. Think Reason 9(3):245–265

Manners-Bell J (2014) Supply chain risk: understanding emerging threats to global supply chains. Kogan Page, London

Marandola A (2012) Ricostruzione "alternativa" del fatto e test di ragionevolezza del "dubbio" in appello. Archivio penale, pp 365–374

Marseglia G, Viola L (2007) La responsabilità penale e civile del medico, Halley ed., Matelica

Marti P, Lanzi P, Bannon L, Sartor G, Contissa G, Masutti A (2011) Liability and Automatation: issues and challenges for socio-technical systems. In: Proceedings SESAR Innovation Days, 29 November-1 December 2011, SESAR WPE, pp 1-5

Marx D (1997) Discipline: the role of rule violation. Ground Eff 2:1–7

Marx D (2001) Patient safety and the 'just culture': a primer for health care executives. Columbia University, New York

Marx D (2009) Whack-a-mole: the price we pay for expecting perfection. Plano, Texas

Masera L (2007) Accertamento alternativo ed evidenza epidemiologica nel diritto penale. Giuffré, Milano, p 166 ff

Masieri C (2016) Accertamento del nesso di causalità nella responsabilità sanitaria: paradigma penale e civile a confronto. Rivista trimestrale diritto processuale civile, pp 1439–1462

Mason K (2017) Fault, causation and responsibility: is tort law just an instrument of corrective justice? In: Freckelton I, Mendelson D (eds) Causation in law and medicine. Taylor & Francis, Ashgate Publishing, Aldershot, pp 147 ff

Massaro A (2017) La legge Balduzzi e la legge Gelli-Bianco sul banco di prova delle questioni di diritto intertemporale alle Sezioni unite l'ardua sentenza. Giurisprudenza penale web, pp 1–8

Mateesco Matte N (1995) La Convenzione di Chicago. Quo vadis OACI? In: Studi in onore di Antonio Lefebvre d'Ovidio, Giuffré, Milano, pp 641 ff

Mauceri F (2015) Al di là di ogni ragionevole dubbio o più probabile che non: note minime sul nesso causale nella responsabilità civile. Jus civile, pp 110–116

McDonald N, Corrigan S, Ward M (2002) Cultural and organizational factors in system safety: good people in bad systems. In: Proceedings of the 2002 International Conference on Human-Computer Interaction in Aeronautics, HCI-Aero, pp 205–209

McGrath J (2015) The Little book of big decision models, Pearson Education Limited, Birmingham City University, Birmingham

Mearns K, Whitaker SM, Flin R (2003) Safety climate, safety management practice and safety performance in offshore environments. Saf Sci 41:641–680

Merry AF (2009) How does the law recognize and deal with medical errors? J Roy Soc Med 102:265–271

Merry AF, McCall Smith A (2001) Errors, medicine and the law. Cambridge University Press, Cambridge

Michaelides-Mateou S, Mateou A (2014) Notification and reporting of aircraft accidents or incidents. In: Pellegrino F (ed) Legislation and regulation of risk management in aviation activity, vol I. Giuffré, Milano, pp 265 ff

Miglio M (2016) In tema di responsabilità medica colposa. Rivista italiana medicina legale e diritto in campo sanitario 3:1250–1256

Milde M (2008) International air law and ICAO. Eleven International Publishing, Utrecht

Milde M (2012) International air law and ICAO. In: Benkö M (ed) Essential air and space law, vol 10. Elven Law, Utrecht, pp 8–10

Miller VB, Jones TL (2011) Creating a just culture: a nurse leader's guide. Hcpro Incorporated, Middleton

Moccheggiani M (2017) Sapere scientifico e ruolo del giudice. Primi appunti. Quaderni costituzionali 3:571–597

Morley FJ, Harris D (2006) Ripples in a pond: an open system model of the evolution of safety culture. Int J Occup Saf Ergon 12(1):3–15

Morreim H (2004) Medical errors: pinning the blame versus blaming the system. In: Sharpe VA (ed) Accountability patient safety and policy reform. Georgetown University Press, Washington, pp 213–232

Morselli E (1990) Analogia e fattispecie penale. L'indice penale, pp 505 ff

Mura A (2017) Attuale insostenibilità dell'epistemologia sottesa alla sentenza Franzese. Cassazione penale 9:3396–3413

Muscolo P (1970) Fondamento, natura e limiti della colpa medica. Giustizia penale 11:449–468

Nefeli Gribaudi M (2012) Consenso e dissenso informati nella prestazione medica. Giuffrè, Milano

Nocco L (2006) Causalità: dalla probabilità logica (nuovamente) alla probabilità statistica, la Cassazione civile fa retromarcia. Danno e responsabilità 12:1239–1245

Nordby K, Helmersen PH, Gilmore D, Johnston N (1995) Do blame and punishment have a role in organisational risk management? Flight Deck 15:33–36

Norman DA (1988) The psychology of everyday things. Basic Book, New York

Norman DA (2013) The design of everyday things. Basic Book, New York

North DM (2000) Let judicial system run its course in crash cases. Aviat Week Space Technol 152 (20):66–67

Nuvolone P (1969a) Colpa civile e colpa penale. In: Nuvolone P (ed) Trent'anni di diritto e procedura penale. Cedam, Padova

Nuvolone P (1969b) Trent'anni di diritto e procedura penale. Cedam, Padova

Oberdiek H (1972) Intention and foresight in criminal law. Mind 81(323):389–400

Page AH (2007) Making just culture a reality: one organization's approach. Agency for Healthcare Research and Quality – AHRQ.gov, in https://psnet.ahrq.gov

Paliero CE (2015) Causalità e probabilità tra diritto penale e medicina legale. Rivista italiana di medicina legale e del diritto in campo sanitario 4:1507–1518

Palma A (2017) Molto rumore per nulla: la Legge Gelli-Bianco di riforma della responsabilità penale del medico. Rivista italiana di medicina legale e diritto in campo sanitario 2:523–541

Palmieri A (2018) In tema di responsabilità colposa per morte o lesioni personali in ambito sanitario. Foro italiano 4:235–238

Palombi E (2018a) Il rispetto del diritto vivente. Rivista penale 4:352–355

Palombi E (2018b) Profili penali della colpa medica nell'evoluzione giurisprudenziale. Archivio giuridico circolazione assicurazione e responsabilità 1:32–39

Pandit MS (2009) Medical negligence: criminal prosecution of medical professionals, importance of medical evidence: some guidelines for medical practitioners. Indian J Urol 25(3):379–383

Panelli S (2015) Inchiesta aeronautica e procedimento penale. In: Pellegrino F (ed) Legislation and regulation of risk management in aviation activity, vol II. Giuffré, Milano, pp 231–237

Panelli S, Scarabello M (2013) Why is it necessary to criminalise negligent behaviour? HindSight 18:58–61

Pardolesi R, Simone R (2017) Nuova responsabilità medica: il dito e la luna (contro i guasti da contatto sociale?). Foro italiano 4:161–175

Patankar MS, Brown JP, Sabin EJ, Bigda-Peyton TG (2012) Safety culture: building and sustaining a cultural change in aviation and healthcare. Saint Louis University (USA), Ashgate, Farnham Surrey, England

Pavich G (2017) La responsabilità penale dell'esercente la professione sanitaria: cosa cambia con la legge Gelli-Bianco. Cassazione penale 7–8:2961–2978

Pellegrino ED (2004) Prevention and medical error: where professional and organizational ethics meet. In: Sharpe VA (ed) Accountability patient safety and policy reform. Georgetown University Press, Washington, pp 83–98

Pellegrino F (2003) Sull'applicabilità dell'Annesso 13 ICAO nell'ordinamento italiano. Diritto dei trasporti, pp 805 ff

Pellegrino F (2007) Sicurezza e prevenzione degli incidenti aeronautici nella normativa internazionale, comunitaria e interna. Giuffré, Milano

Pellegrino F (2013) Just culture principles in Aviation law from a European perspective. Ann Air Space Law, pp 471–490

Pellegrino F (ed) (2014) Legislation and regulation of risk management in aviation activity, vol I. Giuffré, Milano

Pellegrino F (ed) (2015) Legislation and regulation of risk management in aviation activity, vol II. Giuffré, Milano

Pepe J, Cataldo PJ (2011) Manage risk, build a just culture. Health Prog 92(4):56–60

Perez Gonzales JD (2012) ICAO: human factors, management and organization. Aeroscience, pp 13–16

Perleth M, Jakubowski E, Busse R (2001) What is 'best practice' in health care? State of the art and perspectives in improving the effectiveness and efficiency of the European health care systems. Health Policy 56(3):237–238

Perrow C (1984) Normal accidents: living with high risk technologies. Basic Books, New York

Piao CS, Cheng WM, Zhou G (2009) Safety analysis and assessment for human-machine-environment systematic engineering. Ind Saf Environ Prot:44–46

Pidgeon N (1998) Safety culture: key theoretical issues. Work Stress 12(3):202–216

Pidgeon N, O'Leary M (1994) Organisational safety culture: implications for aviation. In: Johnston N, McDonald N, Fuller R (eds) Applications of psychology to the aviation system, Averbury Aviation. Aldershot, Hampshire, pp 21–43

Pidgeon N, O'Leary M (2000) Man-made disasters: why technology and organizations (sometimes) fail. Saf Sci 34:15–30

Piqué F (2017) Il "dolo colpito a mezza via dall'errore": un terreno ancora scivoloso per l'interprete, dove si scontrano esigenze di giustizia sostanziale e necessità di rigorosa applicazione dei presidi garantistici vigenti nel nostro ordinamento. Cassazione penale, pp 1083–1099

Piras P (2018) Un distillato di nomofilachia: l'imperizia lieve intrinseca quale causa di non punibilità del medico. Diritto penale contemporaneo 5:1–11

Pizzi C (2015) The Franzese Judgment: a look at open problems. Paper read at the Conference "Casualty, Counterfactuals and Legal Responsibility", Sassari, 8 October 2015, pp 1 ff. http://www.academia.edu/

Polidori R (2014) Prova indiziaria e giudizio di colpevolezza "oltre ogni ragionevole dubbio". Diritto penale e processo 5:574–585

Pontonio F, Pontonio C (2006) In tema di responsabilità penale del medico: l'art. 2236 c.c. non trova diretta applicazione. Ginecologia Ostetricia Forense 1(1):31–36

Ponzanelli G (2017) Medical malpractice': la legge Bianco-Gelli. Contratto e impresa 2:356–363

Prosdocimi S (1993) Dolus eventualis: il dolo eventuale nella struttura delle fattispecie penali. Giuffré, Milano

Quercietto L (1954) Modifiche alla convenzione di Chicago decise dall'Assemblea dell'O.A.C.I. Rivista aeronautica, pp 1225–1231

Rallo N (2014) ICAO Annex 19: what implications for the state safety programme and for safety oversight? In: Pellegrino F (ed) Legislation and regulation of risk management in aviation activity, vol I. Giuffrè, Milano, pp 101–107

Rasmussen J (1982) Human errors. A taxonomy for describing human malfunction in industrial installations. J Occup Accid 4(2–4):311–333

Rasmussen J (1983) Skills, rules and knowledge: signals, signs and symbols and other distinctions in human performance models. IEEE Trans Sys Man Cybern SMC-13(3):257–266

Rasmussen J, Duncan K, Leplat J (eds) (1987) New technology and human error. John Wiley and Sons, New York

Reason J (1990a) Human error. Cambridge University Press, Cambridge

Reason J (1990b) The contribution of latent human failures to the breakdown of complex systems. Philos Trans Roy Soc Series B Biol Sci, pp 475–484

Reason J (1995) A systems approach to organizational error. Ergonomics 38(8):1708–1721

Reason J (1997) Managing the risks of organizational accidents. Ashgate, Aldershot. e-book published in 2016

Reason J (1998) Achieving a safe culture: theory and practice. Work Stress 3:293–306

Reason J (2000a) Human error: models and management. Br Med J, pp 768–770

Reason J (2000b) Safety paradoxes and safety culture. Injury Control Saf Promot 7(1):3–14

Reinhart RO (1996) Basic flight physiology, 2nd edn. Mcgraw-Hill, New York

Reuter P (2013) Just culture in the real world: flight safety and the realities of society. HindSight 18:49–51

Risicato L (2013) L'attività medica di équipe tra affidamento ed obblighi di controllo reciproco. L'obbligo di vigilare come regola cautelare. Giappichelli, Torino

Risicato L (2018) Le Sezioni unite salvano la rilevanza "in bonam partem" dell'imperizia "lieve" del medico. Giurisprudenza italiana 4:948–954

Rizzo MP (ed) (2009) La gestione del traffico aereo: profili d diritto internazionale, comunitaria e nazionale. Giuffré, Milano

Roncali D (2017) Le linee guida e le buone pratiche: riflessioni medico-legali a margine della legge Gelli-Bianco. Danno e responsabilità 3:280–282

Saponaro L (2018) Il dubbio ragionevole alla ricerca di una definizione. Giurisprudenza italiana 2:469–473

Sardella B (1995) "Eurocontrol" non è un'impresa secondo le norme del diritto comunitario. Giustizia civile 1:12–16

Scarabello M (2017) Work-as-imagined. Work-as-done, and the rule of law. HindSight 25:40–41

Scholtze M (2014) Gestione e mitigazione del rischio nelle attività produttive complesse – Il sistema aviazione. In: Bruni F, Tullio L (eds) Sinistri aeronautici: rischi e responsabilità. Giappichelli, Torino

Schubert F (2013) A just culture in aviation – who is an expert? HindSight 18:46–48

Schwenk W, Schwenk R (1998) Aspects of international cooperation in air traffic management. Martinus Nijhoff, The Hague

Scognamiglio C (2017) Regole di condotta, modelli di responsabilità e risarcimento del danno nella nuova legge sulla responsabilità sanitaria. Corriere giuridico 34(6):740–748

Senders JW, Moray NP (1991) Human error: cause, prediction, and reduction. Lawrence Erlbaum Associates, Hillsdale

Shappell S, Wiegmann D (2001) Applying reason: the human factors analysis and classification system. Human Factors Aerosp Saf 1(1):59–86

Shappell SA (2000) The Human Factors. Analysis and Classification System-HFACS, DOT/FAA/AM-00/7

Shappell SA, Wiegmann D (1997) A human error approach to accident investigation: the taxonomy of unsafe operations. Int J Aviat Psychol 7:269–291

Sharpe CC (1999) Nursing malpractice: liability and risk management. Greenwood publishing Group, Westport

Sharpe VA (2003) Promoting patient safety: an ethical basis for policy deliberation. Hast Center Rep Spec Suppl 33(5):3–18

Sharpe VA (ed) (2004) Accountability patient safety and policy reform. Georgetown University Press, Washington

Skegg P (1998) Criminal prosecutions of negligent health professionals: the New Zealand experience. Med Law Rev 6(2):220–246

Somma E (2014) "Oltre ogni ragionevole dubbio". Una formula enfatica da contestualizzare: meglio, da evitare. Rivista italiana di diritto e procedura penale 1:366–373

Song X, Xie Z (2014) Application of man-machine-environment system engineering in coal mines safety management. In: "2014 ISSST", 2014 International Symposium on Safety Science and Technology, Procedia Engineering 84:87–92

Spadoni AS (2001) Eurocontrol e la funzione unificatrice dello spazio aereo europeo: rafforzamento o svuotamento del principio della sovranità degli Stati. Rivista diritto pubblico e scienze politiche 2:259–267

Spitzmiller R (2011) Selected areas of Italian tort law: cases and materials in a comparative perspective. Fagnano Alto (L'Aquila)

Steel N, Abdelhamid A, Stokes T, Edwards H, Fleetcroft R, Howe A, Qureshi N (2014) A review of clinical practice guidelines found that they were often based on evidence of uncertain relevance to primary care patients. J Clin Epidemiol 67(11):1251–1257

Sternberg R (1996) Cognitive psychology, 2nd edn. Harcourt Brace College Publishers, San Diego

Strachan DMA (1970) The scope and application of the "but for" causal test. Med Legal Regul (MLR) 33:386–389

Strauch B (2004) Investigating human error: incidents, accidents, and complex systems. Ashgate, Aldershot

Strauch B (2017) Investigating human error: incidents, accidents, and complex systems. CRC Press, Taylor & Francis, Boca Raton

Suchman LA (1987) Plans and situated actions: the problem of human-machine communication. Cambridge University Press, Cambridge

Summerton J, Benner B (eds) (2003) Constructing risk and safety in technological practice. Routledge, London

Sumwalt RL (2007) Do you have a safety culture? Flight Safety Foundation, Aero Safety World, pp 38 ff

Terrizzi A (2018) Linee guida e saperi scientifici "interferenti": la Cassazione continua a non applicare la legge Gelli-Bianco. Diritto penale contemporaneo 7:93–112

Thew J (2013) No Fear: How 'Just Culture' is an Antidote to Fear-Based Healthcare. http://www.hl7standards.com

Thomas G (2007) A crime against safety. Air Transp World 44:57–5944

Tirella S (2017) La riforma Gelli è legge. Ecco come cambia la responsabilità medica. Cammino diritto, 28 February 2017

Todeschini N (2017) Approvata la nuova legge sulla responsabilità medica: cosa cambia rispetto alla 'Balduzzi'. Quotidiano giuridico, Ipsoa, Milano

Trovò L (2008) Eurocontrol: in assenza di attività economica non scattano le norme sulla concorrenza. Diritto dei trasporti, pp 163–170

Uttal B (1983) The corporate culture vultures. Fortune Magazine, 17 October 1983

van Asselt MBA, Renn O (2011) Risk governance. J Risk Res 14(4):431–439

van Beuzekom M (2012) Akerboom SP, Hudson PTW, Patient safety in the operating room: an intervention study on latent risk factors. BMC Surg 12(1):10 ff

van Beuzekom M, Akerboom SP, Hudson PTW (2010) Patient safety: latent risk factors. Br J Anaesthesia 105(1):52–59

van het Kaar D (2010) Just culture in civil aviation, not just a fait accompli. Journaal Luchrecht, pp 64 ff

Velliscig L (2018) Assicurazione e "autoassicurazione" nella gestione dei rischi sanitari. Giuffré, Milano

Venturato B (2017) In tema di omicidio colposo, nesso di causalità, colpa. Rivista italiana di medicina legale 4:1596–1610

Vicoli D (2003) Scelte del pubblico ministero nella trattazione delle notizie di reato e art. 112 cost.: un tentativo di razionalizzazione. Rivista italiana di diritto e procedura penale, pp 251–293

Viganò F (2006) Problemi vecchi e nuovi in tema di responsabilità penale per medical malpractice. Corriere del merito 2(8):961–976

Viganò F (2009) Riflessioni sulla c.d. "causalità omissiva" in materia di responsabilità medica. Rivista italiana diritto e procedura penale 4:1679–1725

Viganò F (2013) Il rapporto di causalità nella giurisprudenza penale a dieci anni dalla sentenza Franzese, Relazione all'incontro dibattito presso la Corte di Cassazione il 28 novembre 2012. Diritto penale contemporaneo, n. 3

Von Buri M (1873) Ueber Causalität und deren Verantwortung, Leipzig

Weber L (2017) The Chicago Convention. In: Dempsey PS, Jakhu RS (eds) Routledge handbook of public aviation law. Routledge, Taylor & Francis, Abingdon, Oxon, New York

Weiner EL, Kanki BG, Helmreich RL (eds) (1993) Cockpit resource management. Academic Press, San Diego

Weston I (2014) Benefits of an aviation just culture. In: Pellegrino F (ed) Legislation and regulation of risk management in aviation activity, vol I. Giuffré, Milano, pp 67–72

Westrum R (1993) Cultures with requisite imagination. In: Wise J, Stager P, Hopkin J (eds) Verification and validation in complex man-machine systems. Springer, New York, pp 401–416

Westrum R (1996) Safety of a technological system. In: NTSB Symposium on Corporate Culture and Transportation Safety Proceedings, 24–25 April 1996, Crystal City, Virginia

Whitehead D, Welch Dittman P, McNulty D (2017) Leadership and the advanced practice nurse: the future of a changing health-care environment. FA Davis Company, Philadelphia

Wickens CD, Gordon S, Liu Y (1998) An introduction to human factors engineering. Addison Wesley Longman, New York

Wickens CD, Hollands JG, Banbury S, Parasuraman R (2016) Engineering psychology and human performance, 4th edn. Routledge, Taylor & Francis, London, New York

Wiegmann DA, Shappell SA (2003) A human error approach to aviation accident analysis: the human factors analysis and classification system. Asghgate Publishing Ltd., Aldershot

Wiegmann DA, Zhang H, von Thaden T, Sharma G, Mitchell A (2002) A synthesis of safety culture and safety climate research. Aviation Research Laboratory, University of Illinois, Urbana-Champaign

Wiegmann DA, Zhang H, von Thaden TL, Sharma G, Gibbons AM (2004) Safety culture: an integrative review. Int J Aviat Psychol 14(2):117–134

Winger B, Karner E, Oliphant K (eds) (2018) Essential cases on misconduct. Digest of European tort law, vol 3. The Gruyter, Vienna

Wise J, Stager P, Hopkin J (eds) (1993) Verification and validation in complex man-machine systems. Springer, New York

Woods CP (2017) Operationalizing a just culture policy using a just culture decision guide and toolkit: a DNP. Southeastern Louisiana University, School of Nursing, Hammond

Woods D, Johannesen LJ, Cook RI, Sarter NB (1994) Behind human error: cognitive systems, computers, and hindsight. Crew Systems Ergonomics Information Analysis Center, Wright-Patterson Air Force Base

Woodward S (2017) Rethinking patient safety. Boca Raton, Florida

Yang YZ, Wu LY, Zhang Q (2005) Application of man machine environment system engineering in underground transportation safety. Ind Saf Environ Prot:49–51

Youngberg BJ (2010) Principles of risk management and patient safety. Jones and Bartlett Learning, Sudbury

Youngberg BJ (2012) Patient safety handbook. Jones and Bartlett Learning, Burlington

Yule S (2003) Safety culture and safety climate: a review of the literature. Industrial Psychology Research Centre, University of Aberdeen, King's College, Aberdeen

Zana M (1987) La responsabilità del medico. Rivista critica di diritto privato, pp 159–162

Zana M (2016) Il doppio binario della responsabilità, tra contrattualità ed extracontrattualità. Le novità del disegno di legge Gelli. Le corti fiorentine, pp 19–25

Ziccardi G (2011) Cyber law in Italy. Kluwer, Alphen aan den Rijn

Chapter 2
The Legal Definition of 'Just Culture' in Aviation

2.1 Legal Concept and Definition of "Just Culture" in Aviation

The concept of 'just culture'[1]—elaborated, as we have seen, by James Reason in the 1990s and subsequently developed by other authors and extended to many sectors—has been legally recognized in aviation legislation at international, European and national level.

What are the reasons for the full implementation of a 'just culture' in this sector?

In aviation, there is a need to learn from accidents and incidents through safety investigations so as to take appropriate measures to prevent the repetition of such adverse events.[2]

Therefore, an elaborated regulatory framework has been created as the best way to achieve these objectives.

In order to implement aviation safety regulations it is also very important that an environment exists where minor occurrences are reported voluntarily. To this purpose, it is necessary to create a reporting environment within organisations, regulators and investigation authorities.

Nevertheless, an effective reporting culture depends on how those aviation organisations handle blame and punishment.

In fact, only a very small amount of human actions taken by front-line operators that are unsafe are deliberate and require sanctions of appropriate severity. Aviation personnel should be held accountable only for willful violations and gross negligence.

---

[1]Licu et al. (2013), p. 14 ff.

[2]So 'Just culture', in https://www.skybrary.aero/index.php/Just_Culture.

© Springer Nature Switzerland AG 2019
F. Pellegrino, *The Just Culture Principles in Aviation Law*, Legal Studies in International, European and Comparative Criminal Law 3, https://doi.org/10.1007/978-3-030-23178-1_2

Under "just culture" conditions, on the contrary, individuals are not blamed for 'honest errors'. Therefore, front-line operators should not be punished for reporting errors or making honest mistakes.

In the light of the above, we will try to reconstruct the defining legal framework of a 'just culture' at international, European and national (Italian system) levels.

2.1.1   At International Level

Integration of the 'just culture' concept in aviation at international level originated from a very important technical report, entitled 'A Roadmap to A Just Culture: Enhancing the Safety Environment'. It was prepared in 2004 by the Flight Operations/ATC Operations Safety Information Sharing Working Group of the Global Aviation Information Network (GAIN).[3]

This report refers to a 'just culture', intended to create an atmosphere of trust in which people are encouraged to provide safety-related information, but in which they are also clear about where the line between acceptable and unacceptable behaviour must be drawn.

As shown by this document, the process of clearly distinguishing between acceptable and unacceptable behaviour takes place in a collaborative environment of trust and puts together "different members of an organization that might often have infrequent contact in policy decision making.[4] This contact, as well as the resulting common understanding of where the lines are drawn for punitive actions, enhances the trust that is at the core of developing Just Culture".[5]

At the regulatory level, the first discussions and findings on a 'just culture' took place in 2007, during the 36th International Civil Aviation Organization (ICAO)[6] Assembly.[7]

In fact, the Working Paper presented in the 36th General Assembly session, entitled "Implementation of a 'just culture' concept",[8] stressed that its purpose was "to support the implementation of an adequate 'Just Culture' concept in order to address the need for the protection of safety reporting and sharing of information while respecting the principles of administration of justice and freedom of information".[9]

---

[3]GAIN, A Roadmap to a Just Culture: Enhancing the Safety Environment, cited above, p. 4 ff.

[4]Orasanu (1993), p. 137 ff.; Weiner et al. (1993); Murray (1997), 7, p. 83 ff.; Hudson (2002), p. 74.

[5]From GAIN, A Roadmap to a Just Culture, cited above, IX.

[6]Giannini (1952), p. 579 ff.; Monaco (1981), p. 403; Sciolla Lagrange (1990), p. 1 ff.

[7]ICAO Assembly 36th Session, Montréal, 18–28 September 2007.

[8]ICAO Working Paper Assembly – 36th Session, Technical Commission, A36-WP/232 TE/76 of 18 September 2007, Agenda Item 28 entitled "Implementation of a 'just culture' concept", supra.

[9]See Working Paper of I CAO Assembly, 36th Session, Technical Commission, A36-WP/232, TE/76, supra, p. 1.

Consequently, according to the ICAO Assembly Resolution A36-7,[10] adopted in the same Assembly session, the achievement of the 'just culture' objectives requires the setting up of "an environment in which the reporting and sharing of information is encouraged and facilitated and in which remedial action is undertaken in a timely fashion when deficiencies are reported".[11]

This means that an electronic occurrence reporting system[12] has to be established to ensure that errors are reported and analysed and appropriate follow-up action is taken.

The underlying rationale of this provision is clear: people are less willing to inform the organization where they work about their own errors if they fear negative consequences in terms of prosecution by disciplinary or judicial authorities.

In fact, it is natural that operators[13] may be less willing to report their errors and other safety issues if they are afraid of prosecution or punishment. For example, consider the controller who personally reported that his tiredness caused a lack of vigilance, which led him to give a wrong landing clearance.

On the same occasion of the 36th General Assembly, the ICAO recognized a definition of the 'just culture' concept already elaborated at the European level, in the following terms: "A culture in which front line operators or others are not punished for actions, omissions or decisions taken by them that are commensurate with their experience and training, but where gross negligence, wilful violations and destructive acts are not tolerated".[14]

Therefore, ICAO did not elaborate this definition but expressly recognized and supported it with the purpose of implementing an adequate 'just culture' concept.

This evolutionary process that contributed to defining the 'just culture' concept at the international level has been promoted by Eurocontrol,[15] considering that to

---

[10]A36-7 ICAO Global Planning for Safety and Efficiency, in Resolutions adopted by the Assembly, Assembly – 36th Session, Montréal, 18–28 September 2007, provisional edition, September 2007.

[11]Resolution A36-7, ICAO Global Planning for Safeguard Efficiency, Appendix A (Global Aviation Safety Plan).

[12]O'Leary and Chappel (1997), p. 11 ff.

[13]See Vaughan (1999), p. 271 ff.; Cohen-Charash and Spector (2001), p. 278 ff.

[14]ICAO Working Paper Assembly – 36th Session, Technical Commission, A36-WP/232 TE/76 of 18 September 2007, Agenda Item 28 entitled "Implementation of a 'just culture' concept", presented by Portugal, on behalf of the European Community and its Member States, by the other States Members of the European Civil Aviation Conference and by Eurocontrol, par. 1 (development of an adequate "just culture" concept), para. 1.4, 2 and Appendix letter a), 4.

[15]International organization established by Convention relating to Co-operation for the Safety of Air Navigation of 13 December 1960, as amended by the amending Protocol of 12 February 1981 (amended Convention) and by the Brussels Protocol of 27 June 1997 (revised Convention), consolidating the Eurocontrol International Convention relating to Co-operation for the Safety of Air Navigation of 13 December 1960. See Sardella (1995), p. 12 ff.; Schwenk and Schwenk (1998), p. 129; Spadoni (2001), p. 259 ff.; Trovò (2008), p. 163 ff.

encourage incident reporting and information sharing[16] is essential in order to improve aviation safety.[17]

The Eurocontrol technical document, called "Just Culture Guidance Material for Interfacing with the Judicial System", mentioned above, defined the 'just culture' as "a culture in which front line operators or others are not punished for actions, omissions or decisions taken by them that are commensurate with their experience and training, but where gross negligence, willful violations and destructive acts are not tolerated".

Additionally, in the same document it has been noted that "a Just Culture is not just another safety-related initiative" but "it is the only way to proceed towards enhancing safety problems at the heart of a Just Culture in different ways".

It is clear that the definition of 'just culture' contained therein fully reflects the text recognized by the ICAO.

The mirror definitions, laid down by the ICAO Assembly and the Eurocontrol Guidance, respectively, show a clear dividing line between non-punishable (acceptable or proper) behaviour and punishable (unacceptable or improper) behaviour.[18] The latter refers to cases of 'willful misconduct' (intentional fault), 'gross negligence', destructive acts, severe and serious disregard of an obvious risk and/or profound failure of professional responsibility, while the former (acceptable behaviour) relates to the conduct that is in line with people's experience and training, namely the so-called 'ordinary negligence' (innocent, honest, excusable error).[19]

This classification is also in line with another document, prepared by Eurocontrol, entitled "Establishment of 'Just Culture' Principles in ATM Safety Data Reporting and Assessment",[20] which states as follows "Gross negligence and criminal offences

---

[16]International organization established by Convention relating to Co-operation for the Safety of Air Navigation of 13 December 1960, as amended by the amending Protocol of 12 February 1981 (amended Convention) and by the Brussels Protocol of 27 June 1997 (revised Convention), consolidating the Eurocontrol International Convention relating to Co-operation for the Safety of Air Navigation of 13 December 1960; Sardella (1995), p. 12 ff.; Schwenk and Schwenk (1998), p. 129; Spadoni (2001), p. 259 ff.; Trovò (2008), p. 163 ff.

[17]See Licu and van Dam (2013), p. 18 ff. and Licu and van Dam (2014), p. 55 ff.

[18]See Starrantino and Finocchiaro (2014), p. 27.

[19]You can find a graphic representation of these differences in SwissATCA, Just Culture Manual for ATCO, ANSE & ATSEP, Behavior after an incident and further proceedings, p. 8, in https://skybrary.aero/bookshelf/books/4222.pdf. See Muchinsky (1997); Moorman and Pipers (2002).

[20]Eurocontrol, ESARR Advisory Material/Guidance Document (EAM 2/GUI 6) of 31 March 2006, ed. 1, p. 17. This guidance material is derived from the research of existing best practices and in particular from the conclusions and recommendations of Eurocontrol's "Safety Data Reporting and Data Flow Task Force" report. This document is intended to be supporting guidance for anyone involved in implementing ESARR 2, especially when encountering difficulty in introducing reporting and assessment systems. ESARR is acronym for Eurocontrol Safety Regulatory Requirements. These requirements contain technical rules on safety, which need be implemented in the European Union or national regulatory framework of its Member States. ESARR 2 is entitled "Reporting and Assessment of Safety Occurrences in ATM", edition 3, 2 December 2009. It defines a list of ATM-related occurrences which shall, as a minimum, be reported and assessed by States (Appendix A), defines the minimum appropriate safety data which shall be collected and reported to

are well defined – by nature they are deliberate acts. Omissions, slips, lapses, mistakes and violations fall under the category of honest mistakes. However, even though there is no strict dividing line between these two major categories (deliberate on one side, unintentional on the other), it is necessary to investigate each event to determine into which category it will be placed".

In other words, a 'just culture' protects operators from being at fault for 'honest mistakes', but does not protect them in case of intentional fault or 'gross negligence'. Therefore, someone who consciously takes an irresponsible risk should be sanctioned.

As mentioned before, error is mostly regarded as an unintentional deviation between what was actually done and what should have been done, while a 'willful violation' is a deliberate act or omission to deviate from established procedures, protocols, norms or practices.[21]

In March 2006, following ICAO Assembly resolutions A33-17[22] and A35-17[23] on protection of accident, incident and other safety data, the ICAO Council adopted the document entitled "Legal guidance for the protection of information from safety data collection and processing systems (SDCPS)".[24] The effective applicability date of this document was 23 November 2006 through a new Attachment, letter E,[25] to ICAO Annex 13, which contains general principles and offers more specific guidance.

This Attachment states that safety information should be protected from inappropriate use, for purposes different from that for which it was collected (*i.e.* for disciplinary, civil, administrative and criminal proceedings against operational personnel and/or disclosure of information to the public). This objective must not interfere with the proper administration of justice.

In fact, the same Attachment holds: 'The sole purpose of protecting safety information from inappropriate use is to ensure its continued availability so that proper and timely preventative actions can be taken and aviation safety improved. It is not the purpose of protecting safety information to interfere with the proper administration of justice in States.'

Eurocontrol by States, expressed in terms of high-level safety indicators (Appendix B) and includes a glossary of terms with harmonised definitions (Appendix C). The majority of ESARR 2 requirements has been transposed into European Community Law by Directive 94/56/EC and Regulation (EU) No. 376/2014 (see below). See Nastro (1993), p. 407 f.; Cook (2016), para. 8.3.2 (Safety, Transparency and 'Just Culture'); Brooker (2007), p. 3 ff.; Van Antwerpen (2009), p. 83; Pellegrino (2009), p. 216, Pellegrino (2012, 2014); Bieder and Bourrier (2017), para. 1 ff.

[21]So ICAO Safety Management Manual (SMM) Doc 9859 AN/474, Third Edition, 2013, 2.5, pp. 2–8.

[22]A33-17 "Non-disclosure of certain accident and incident records", in Resolutions adopted at the 33rd session of the Assembly, provisional edition, 2001, p. 69.

[23]A35-17 "Protecting information from safety data collection and processing systems in order to improve aviation safety", Resolutions adopted by the Assembly, provisional edition, October 2004, p. 71.

[24]See Espinola et al. (2005), p. 26.

[25]Entitled "Legal guidance for the protection of information from safety data collection and processing systems", then transposed in Attachment B to ICAO Annex 19, ATT B-1, see below.

The outcome of the 2007 General Assembly,[26] together with the conclusions of the 2008 AIG Divisional meeting,[27] the recommendations of the ICAO HLSC Conference 2010[28] and the Safety Information Protection Task Force (SIPTF)[29] converged in resolutions A37-2 ("Non-disclosure of certain accident and incident records")[30] and A37-3 ("Protecting information from safety data collection and processing")[31] of the 37th session of the ICAO Assembly.[32]

In the first Resolution (A37-2), *inter alia*, the Assembly requested the Council to improve the provisions on the protection of certain accident and incident records, with the aim of facilitating the implementation of Annex 13 provisions on the protection of safety information (para. 5.12), taking into account "the necessary interaction between safety and judicial authorities in the context of open reporting culture".

The second Resolution (A37-3) stressed the importance of a balanced environment in which disciplinary measures are not taken against operational personnel for their "actions that are commensurate with their experience and training, but where gross negligence and willful violations are not tolerated". The Assembly's intention to firmly reiterate the contents of the 'just culture' definition, recognized in the 36th session, is evident.

Both resolutions advised the Council to strike a balance between the need for the protection of safety information and the interest for the proper administration of justice.[33]

These resolutions, however, did not contain a definition of 'just culture', thus showing that the international civil aviation community wanted to maintain the previous wording,[34] on which all agreed.

---

[26]ICAO General Assembly, 36th Session, Montréal, 18–28 September 2007.

[27]AMC/MA/Accident Investigation and Prevention (AIG) Divisional Meeting, 13–18 October 2008, Montréal, Canada, that, *inter alia*, highlighted the absence of a voluntary reporting system and the absence of safety data analysis by the State aviation authorities. See Working paper AIG/08-WP/20 of 25 June 2008.

[28]See the Working paper HLSC 2010-WP/8, 31 March 2010, of the High-Level Safety Conference 2010, Montréal, 29 March–1 April 2010, on the theme "Towards the proactive management of safety", topic "Implementing new safety management process".

[29]In its final report the SIPTF recommended close cooperation between safety (in particular, just culture) and justice. See, in particular, SIPTF fourth Meeting on Just Culture, Montréal, 21–25 January 2013, presented by Roderick van Dam, SIPTF/4-IP/2 of 17 January 2013.

[30]Published in Resolutions adopted by the Assembly, November 2010 (provisional edition).

[31]Published in Resolutions adopted by the Assembly, November 2010 (provisional edition).

[32]ICAO General Assembly 37th session, Montréal, 28 September–8 October 2010 resolutions A37-2 (Non-disclosure of certain accident and incident records) and A37-3 (Protecting information from safety data collection and processing system in order to improve aviation safety), published on November 2010 (provisional edition).

[33]See Licu and van Dam (2013), p. 18 ff. and Licu and van Dam (2014), p. 55 ff.

[34]As follows: "a culture in which front-line operators or others are not punished for actions, omissions or decisions taken by them that are commensurate with their experience and training, but where gross negligence willful violations and destructive acts are not tolerated".

Finally, although in chapter 1 of ICAO Annex 19 (Safety Management) there is no a legal definition of 'just culture', nevertheless it contains a provision that is fully consistent with the approach taken in the 36th General Assembly.

In fact, Article 5(3.2),[35] providing a recommended practice[36] with regard to safety reporting, is integrated by a note as follows: "A reporting environment where employees and operational personnel may trust that their actions or omissions that are commensurate with their training and experience will not be punished is fundamental to safety reporting".

In this very important Annex, however, ICAO makes its preference to the concept of 'aviation safety', defined as "The state in which risk associated with aviation activities, related to, or in direct support of the operation of aircraft, are reduced and controlled to an acceptable level".[37]

This means that any action that significantly and seriously affects aviation safety, lowering its level and posing an unacceptable risk is not tolerated. Therefore, it requires corrective and preventive measures.

But linking the two provisions contained in Annex 19, mentioned above, leads us to conclude that any action of aviation personnel commensurate with their experience and training, even if it slightly affects aviation safety, on the contrary, is considered acceptable and then tolerated.

Actually, the 38th ICAO Assembly[38] led the ICAO Council to take appropriate measures to ensure that provisions contained in international standards and recommended practices (SARPs)[39] of Annex 19 and guidance materials on the protection of information collected through safety data collection and processing systems (SDCPS) are reinforced. This was done with a view to sustain the availability of safety information, required for the management, maintenance and

---

[35] Article 5(3.2): "States should extend the protection referred to in 5.3.1 to safety data captured by, and safety information derived from, mandatory safety reporting system and related sources".

[36] Defined as "any specification for physical characteristics, configuration, material, performance, personnel or procedure, the uniform application of which is recognised as desirable in the interest of safety, regularity or efficiency of international air navigation, and to which Contracting States should endeavour to conform in accordance with the Convention" (so ICAO Assembly Resolution A36-13, Appendix A. ICAO Doc 9902, Assembly Resolutions in force as of 28 September 2007, in http://www.icao.int/icaonet/dcs/9902/index.html.

[37] Chapter I of the ICAO Annex 19. In addition, the ICAO "Universal Safety Oversight Audit Programme Continuous Monitoring Manual" contains the following similar definition of safety: "The state in which the possibility of harm to persons or property damage is reduced to, and maintained at or below, an acceptable level through a continuing process of hazard identification and risk management". See Di Renzo (2015), p. 155 ff.; Di Carlo (2015), p. 203 ff.; Guarrera (2015), p. 171 ff.; Bernabei and Barbafina (2015), p. 163 ff.

[38] ICAO General Assembly 37th session, Montréal, 24 September–4 October 2013.

[39] Standards and Recommended Practices (SARPs) are technical specifications adopted by the Council of ICAO in order to achieve "the highest practicable degree of uniformity in regulations, standards, procedures and organization in relation to aircraft, personnel, airways and auxiliary services in all matters in which such uniformity will facilitate and improve air navigation" (Article 37 of the Chicago Convention).

improvement of safety, taking into account the necessary interaction between safety and judicial authorities,[40] in the context of an open reporting culture.[41]

The need of a 'just culture' and a policy for the protection of safety data is mentioned in another contemporary ICAO document, the 'Safety Management Manual' (SMM),[42] referring to the risk that aviation personnel and service providers could be "exposed to unfair or inappropriate/disciplinary/judicial proceedings" as a result of occurrence reporting (Article 5(3.14)).

'Safety culture' and the need to protect safety data and information have been prominent topics at the second High Level Safety Conference, held in 2015,[43] where it was clarified that "These protections are essential for the ongoing availability of safety data and safety information, and forms the basis of a just culture".[44]

This Conference emphasised the need to adopt new and enhanced provisions on the protection of accident and incident records and information, collected for the purpose of maintaining and improving safety. Another point of discussion was about how to support Contracting States in implementing this new legal framework through guidance material, seminars, workshops, and so on. In turn, States were called upon to undertake further action and legal adjustments to efficiently implement the new protective framework.

On 1st October 2015 the European Commission hosted the signing of the "European Corporate Just Culture Declaration".[45] This document, although is a non-legally binding instrument, committed the signatories (a large number of aviation stakeholders) to implementing 'just culture'. It constitutes a set of key principles that each organisation is encouraged to implement in the context of its 'just culture' internal rules in support of the new occurrence reporting Regulation No. 376/2014.

The following principles contained in this Declaration are very important: "acting safely is a top priority", "when assessing individual responsibility, organisations should focus on determining if actions, omissions or decisions taken were commensurate with experience and training, and not on the outcome of an event", "analysis of reported occurrences by organisations should focus on system performance and contributing factors first and not on apportioning blame and/or focus on individual responsibilities", "organisations should promote effective implementation of Just Culture principles within the organisation at all levels and with all parties, including

---

[40]Licu et al. (2013), p. 15.

[41]See ICAO Assembly – 38th session, Technical Commission, Draft text for the Report on Agenda Item 27, A38-WP/377 TE/167 of 26 September 2013, Article 3, p. 27-9, that supersedes Resolution A37-3.

[42]Doc 9859 AN/474, Third Edition, 2013.

[43]The Second High-level Safety Conference (HLSC2015) held from 2 to 5 February 2015 at ICAO Headquarter, in Montréal, Canada.

[44]So Gilberto Lopez Meyer, IATA Senior President, Safety and Flight Operations. At the same Conference he stressed: "It is only natural that people and organizations would be less willing to report their errors and other safety issues if they are afraid of punishment or even prosecution".

[45]https://ec.europa.eu/transport/sites/transport/files/modes/air/events/doc/2015-10-01-just-culture/declaration.pdf.

their representatives", "Just Culture internal rules should include, amongst others, the definition of a process, including the actors involved, to determine an unacceptable behaviour", "to effectively implement a Just Culture staff at all levels, as well as top management, should understand and accept their responsibility with regards to Just Culture principles and internal rules and their promotion".

Subsequently, on 2 March 2016, the safety management SARPs[46] were strengthened through the approval of the first amendment to Annex 19, mentioned above.

Amendment 1, which will be implemented from November 2019, enhances legal safeguards intended to ensure the appropriate use and protection of safety information and to support proactive safety improvement strategies.

Inter alia, it introduces references to the 'safety culture' in Annex 19 in order to enhance protection of safety data and safety information, as well as their sources.[47] As we have seen, 'safety culture' is a broader concept of which 'just culture' is part. The latter enables the former to exist.

The importance of the implementation of 'just culture' principles in aviation law is demonstrated by the recent working paper, entitled 'Improving Just Culture',[48] presented within the 39th ICAO Assembly session by the Civil Air Navigation Services Organisation (CANSO).

According to the CANSO concept of 'just culture', this is an "atmosphere in which people are encouraged for providing essential safety-related information, but in which they are also clear about where the line must be drawn between acceptable and unacceptable behaviour".

The same Organization's working paper says that a healthy 'just culture' "plays a vital role in a successful safety culture by encouraging employees to report safety incidents and hazardous conditions" and "enables the proactive identification of safety trends. A critical part of Just Culture is also the responsibility to be consistently intolerant of willful misconduct or reckless behaviour. Adopting a clearly

---

[46]"The safety management SARPs are intended to assist States in managing aviation safety risks, in coordination with their Service Providers. Given the increasing complexity of the global air transportation system and its interrelated aviation activities required to assure the safe operation of aircraft, the safety management provisions support the continued evolution of a proactive strategy to improve safety performance. The foundation of this proactive safety strategy is based on the implementation of a State safety programme (SSP) that systematically addresses safety risks, in agreement with the implementation of the safety management systems (SMS) by the service providers" (so www.icao.int/safety/safetymanagement/pages/sarps.aspx).

[47]The second edition of Annex 19 is comprised of the following: an upgrade of State Safety Programme (SSP) provisions; an enhancement of the Safety Management System (SMS) provisions; an upgrade of provisions for the protection of safety data, safety information and related sources.

[48]ICAO Working Paper, Assembly – 39th Session. Technical Commission 'Agenda Item 36: Aviation safety and air navigation implementation support "Improving Just Culture" (presented by the Civil Air Navigation Services Organisation CANSO), A39-WP193 TE/73 of 25 August 2016. See Antonini and Franchi (2005).

defined Just Culture policy and programme will benefit the safety management of each aviation organization".[49]

The CANSO's definition of 'just culture', despite a different wording, does not conceptually deviate from that recognised by the EU aviation legislation and reproduces almost entirely the Reason's definition.[50]

2.1.2   At European Level

Since the 2000s, the European Union enacted 'just culture' in its legal order.

In fact, in 2003 the Council of the European Union, together with the European Parliament, adopted the above mentioned Directive 2003/42/EC on occurrence reporting in civil aviation.

Under Article 8(3) of this Directive, without prejudice to the applicable national rules of criminal law, Member States must refrain from instituting proceedings and prosecuting aviation operators in respect to "unpremeditated or inadvertent infringements of the law which come to their attention only because they have been reported under the national mandatory occurrence-reporting scheme, except in cases of gross negligence".

This provision has been elaborated with the intent to resolve the lack of uniformity of the Member States' national laws.

In the light of this concept, no prosecution is allowed within the European Union against actions, omissions or decisions of the reporters who behaved as a reasonable person, in case of 'unpremeditated' or 'inadvertent' infringement of the law.

'Unpremeditated' violation refers to unintentional conduct, which is not characterized by willful intent and is not planned in advance, whereas 'inadvertent' refers to an unconscious infringement. Both these behaviours have to be tolerated.

On the contrary, national jurisdiction will remain unaffected in respect to the prosecution in the cases involving intentional wrongdoing or 'gross negligence'.

On December 2007, a few months after the adoption of the 36th ICAO Assembly Resolution mentioned above, the European Commission adopted the Communication entitled 'First Report on the implementation of the Single Sky Legislation: achievements and the way forward'.[51] This is the first report on the progress of

---

[49]McDonald et al. (2000), p. 151 ff. see; Corrigan (2003).

[50]Reason (1997), p. 195. See also Harris (2013), p. 91 ff.

[51]'First Report on the implementation of the Single Sky Legislation: achievements and the way forward': Communication COM (2007) 845 final, Brussels, 20 December 2007. This report contains an evaluation of the results achieved in implementing the Single Sky legislation, including information about developments in the sector, in the light of the original objectives and with a view to future needs for future development of the SES.

implementation,[52] which lays down the Commission's views on the need for future development of the Single Sky (SES).[53]

Article 12(2) of the Single Sky Framework Regulation (No. 549/2004)[54] required the Commission to review the application of relative legislation and report periodically to the European Parliament and Council about it.

Although this report pays attention to the 'just culture' from a legal point of view, it does not contain a definition of this concept, while implicitly referring to that accepted by ICAO.

This 'soft law' document, *inter alia,* recommends[55] that Member States apply safety management principles consistently, in particular, to facilitate the uniform application of a 'just culture'.

Along the same line, Regulation (EC) No. 216 of 20 February 2008[56] on common rules in the field of civil aviation (so-called "Framework Regulation"), although not introducing the definition of 'just culture', emphasizes the need to make a distinction between honest and grossly negligent behaviours.

In this respect, Article 16(2), states as follows:

> Without prejudice to applicable rules of criminal law, Member States shall refrain from instituting proceedings in respect of unpremeditated or unintentional infringements of the law which come to their attention only because they have been reported pursuant to this Regulation and its implementing rules. This rule shall not apply in cases of gross negligence.

Subsequently, the same definition of 'just culture' already contained in the ICAO documents mentioned above, has been fully transposed into European legislation.

---

[52]Stadler (2009), p. 267.

[53]See Valente (2005), p. 203 ff.; Masala and Rosafio (2006); Vincenzi (2010), p. 316 ff.; Calleja Crespo and Mendes de Leon (2011); Trovò (2011), p. 24 ff.; Petrick-Felber (2014), p. 12 ff.; Cook (2016), para. 8.3.2 (Safety, Transparency and 'Just Culture').

[54]Regulation (EC) No. 549/2004 of the European Parliament and of the Council of the 10th March 2004 laying down the framework for the creation of the Single European Sky (SES) (O.J. L 96 of 31 March 2004). For a commentary, see Turco Bulgherini (2009), p. 338; Rizzo (2006), p. 407; Xerri (2012), p. 67 ff.

[55]Recommendation No. 8 of the High Level Group (HLG), contained in the Annex to Communication COM (207)845 final of 20 December 2007, cited above.

[56]Regulation (EC) No. 216/2008 of the European Parliament and of the Council of 20 February 2008 on common rules in the field of civil aviation and establishing a European Aviation Agency and repealing Council Directive 91/670/EEC, Regulation (EC) No. 1592/2002 and Directive 2004/36/EC (O.J. L 79 of 19 March 2008). The scope of this Regulation, initially limited to airworthiness, was then extended to the flight operations and subsequently to airport safety and to air traffic management and control systems. On December 2015 a proposal for a Regulation of the European Parliament and of the Council on common rules in the field of civil aviation and establishing a European Union Aviation Safety Agency and repealing Regulation (EC) No. 216/2008 has been presented by the European Commission. See also Regulation (EC) No. 1108/2009 of the European Parliament and of the Council of 21 October 2009 amending Regulation (EC) No. 216/2008 in the field of aerodromes, air traffic management and air navigation services and repealing Directive 2006/23/EC (O.J. L 309 of 24 November 2009). For a commentary, see Coman-Kund (2018), notes 57–58; Cassatella (2017), p. 25.

In fact, the Commission Regulation (EU) No. 691/2010,[57] laying down a performance scheme[58] for air navigation services, at Article 2, letter k), likewise states: "Just culture' means a culture in which front line operators or others persons are not punished for actions, omissions or decisions taken by them that are commensurate with their experience and training, but where gross negligence, willful violations and destructive acts are not tolerated".

Furthermore, this Regulation provides for Key Safety Performance Indicators (KPIs)[59] and binding targets on the key performance areas of safety, environment, capacity and cost-efficiency.[60]

'Key performance indicators' means the performance indicators used for the purpose of performance target setting, while 'binding target' means a performance target adopted by Member States as part of a national or functional airspace block (FAB)[61] performance plan (Article 2, points d) and f) of Regulation No. 691/2010).

According to Annex 1[62] to this Regulation, the third national/FAB safety KPI shall be the reporting by Member States, through a dedicated *questionnaire*, which measures the level of 'just culture'.

Consequently, in the context of SES (Single European Sky),[63] the measurement of 'just culture' can also be regarded as a safety performance indicator, one of the pillars[64] of a good 'safety culture', which reflects the real commitment, at all levels, to safety.

---

[57] Commission Regulation (EU) No. 691/2010 of 29 July 2010 "laying down a performance scheme for air navigation services and network functions and amending Regulation (EC) No. 2096/2005 laying down common requirements for the provision of air navigation services" (O.J. L 201 3 August 2010). It has been updated by Commission Implementing Regulation (EU) No. 390/2013 of 3 May 2013, laying down a performance scheme for air navigation services and network functions (O.J. L 128 of 9 May 2013). For a commentary see Calleja Crespo and Mendes de Leon (2011), p. 281; Rossi dal Pozzo (2014), p. 22; Togan (2016), p. 224. For critical considerations about this definition, see Vernizzi (2017), p. 266 f.

[58] The establishment of these schemes derives from the so-called "Framework Regulation", mentioned above.

[59] The other two KPIs are: effectiveness of Safety Management and application of the severity classification of the Risk Analysis Tool to Separation Minima Infringements. For safety culture indicators, applied in the domain of nuclear safety, see Reiman and Pietikäinen (2010).To pave the way forward in establishing a set of KPIs for Safety in ATM, Eurocontrol established a 'Safety Data Reporting and Data Flow Task Force' (SAFREP).

[60] According to the KPI definition contained in Article 2, letter d), of the Regulation No. 691/2010.

[61] A Functional Airspace Block (FAB) is defined in the Single European Sky legislative package namely Regulation (EC) No. 1070/2009 of 21 October 2009 (O.J. L 300 of 14 November 2009) amending Regulation (EC) No. 549/2004 (Framework Regulation), as an airspace block based on operational requirements and established regardless of State boundaries, where the provision of air navigation services and related functions is optimized through enhanced cooperation among air navigation service providers (ANSPs) or, when appropriate, an integrated provider, always in a performance-driven perspective. See Trovò (2011), p. 24 ff.; Bufo (2012), p. 165 f.

[62] Annex 1, Section 2, "For national or Functional Airspace Block (FAB) target setting", letter c).

[63] See Vincenzi (2010), p. 310 ff.; Xerri (2012), p. 67 ff.; Tytgat (2012), p. 93 ff.; Preti (2012), p. 113 ff.

[64] The other two indicators specified in Regulation No. 1216/2011 (point 2) are: the effectiveness of safety management and the application of severity assessment of the Risk Analysis Tool.

As specified from the Commission Implementing Regulation (EU) No. 1216/2011,[65] these safety indicators[66] should be further developed jointly by the Commission, the Member States, the European Aviation Safety Agency (EASA)[67] and Eurocontrol.[68]

Other subsequent European regulations reiterated the common (international and European) definition of 'just culture'.

The drafters of the well-known Regulation No. 996/2010[69] on the investigation and prevention of accidents and incidents, without feeling the need to redefine a 'just culture', yet in accordance with the above mentioned ICAO Resolutions, stressed that: "The civil aviation system should equally promote a non punitive environment facilitating the spontaneous reporting of occurrences and thereby advancing the principle of 'just culture' (recital 24).

In the light of this provision, spontaneous reporting is considered to be a means of ensuring the application of 'just culture' principles, in an atmosphere of complete trust, without the fear of punishment as a consequence of this reporting activity, and allowing the civil aviation system to learn from past defects or other abnormal events.[70]

In this framework, 'just culture' is the condition that promotes a better working environment by holding operators 'accountable', without making them responsible or liable and consequentially without punishing them.

Considering that the provision under examination aims at making aviation operators accountable for their acts in a transparent manner, it needs to clarify the meaning of the concept of 'accountability'.[71] In legal terms, this is not synonymous with responsibility or liability, although these expressions are often used interchangeably in common language.

'Responsibility' is legally understood as an obligation to perform a particular task, whereas 'liability'[72] means the subjection to legal consequences if one's duties

---

[65]Commission Implementing Regulation (EU) No. 1216/2011 amending Commission Regulation (EU) No. 691/2010 laying down a performance scheme for air navigation services and network functions (O.J. L 310 of 25 November 2011, 3).

[66]Hudson (2009), 47, p. 483 ff.

[67]Randazzo (2004), p. 847 ff.; Giemulla and Weber (2011), p. 321 f.; Marino (2013), p. 45 ff. See also European Flight and Aviation Safety Regulations Handbook, vol. I, System, procedures and important regulations, USA International Business Publications, Washington, 2009, 34; Pifisterer (2017), p. 282 ff.

[68]With the aim of developing these safety indicators, documents such as EASA Safety Plan and Eurocontrol's Risk Analysis Tool and Safety Framework Maturity Survey should be taken into account.

[69]Regulation (EU) No. 996/2010 of the European Parliament and of the Council of 20 October 2010 on the investigation and prevention of accidents and incidents in civil aviation and repealing Directive 94/56/EC (O.J. L 295 12/11/2010, 35). For a commentary, see Franchi (2012), p. 41 ff.

[70]See Michaelides-Mateou and Mateou (2014), p. 271 ff.; Baumgartner and Schorer (2017), p. 3.

[71]See Dekker (2012), p XI f. See also Lerner and Tetlock (1999), p. 255 ff.

[72]De Franchis (1985), p. 1402.

are not fulfilled. The term 'accountability'[73] comes from 'account' which means a report or description of a past event or experience. Therefore, 'accountability' literally means the ability to report and to account for events or experiences that have already happened, whereas, in legal terms, it refers to the state or fact of being accountable and answerable for your actions and resulting consequences (answerability).

EU Regulation No. 376/2014, defining the reporting system, repealed Directive 2003/42/EC. It reiterates the same definition of 'just culture' already contained in Regulation No. 691/2010.

In fact, Article 2(12), states that

> 'Just culture' means a culture in which front line operators or other persons are not punished for actions, omissions or decisions taken by them that are commensurate with their experience and training, but where gross negligence, willful violations and destructive acts are not tolerated.

In addition, many other provisions of this Regulation are expressly referred to as 'just culture'.

In particular, attention must be drawn to recital 34 that, in order to ensure protection of occurrence reports, reads as follows: "The internal 'just culture' rules adopted by organisations pursuant to this Regulation should contribute in particular to the achievement of this objective".

Consistently, recital 36 adds that "the civil aviation system should promote a 'safety culture' facilitating the spontaneous reporting of occurrences and thereby advancing the principle of a 'just culture'. An environment embracing 'safety culture' principles "should not prevent action being taken where necessary to maintain or improve the level of aviation safety".

In addition, the last recital highlights that there is a close link between a 'just culture' and a 'safety culture' where says: 'Just culture' is an essential element of a broader 'safety culture', which forms the basis of a robust safety management system".

In fact, as we have seen, a 'just culture' is an essential element of a broader 'safety culture', which in turn forms the basis of a robust Safety Management System (SMS).[74]

---

[73]For a definition see Roughton and Mercurio (2002), p. 57; Sharpe (2015), p. 185 ff.; Castellani (2015), p. 80.

[74]According to the ICAO definition, Safety Management System (SMS) is "a systematic approach to managing safety, including the necessary organisational structures, accountabilities, policies and procedures" (Doc 9859, AN/474, *Safety* Management Manual (SMM), third edition, 2013, xii, Definitions). According to the Eurocontrol definition, Safety Management System (SMS) is "a systematic and explicit approach defining the activities by which safety management is undertaken by an organisation in order to achieve acceptable or tolerable safety" (ESARR 3, Use of Safety Management Systems by ATM Service Providers, edition 1, Bruxelles, 17 July 2000, Appendix A, Glossary - Terms and Definitions, 16). See Ludwig (2009), Stolzer et al. (2011) and Stolzer and Goglia (2015).

According to the close link existing between 'just culture' and 'safety culture', recital 36 should be read in conjunction with recital 37, which states that a 'just culture' should encourage individuals to report safety-related information.

This means that a 'just culture' requires an environment of trust between regulators, law enforcement, investigation authorities, organization and judiciary. An atmosphere in which occurrence reporting are encouraged and this allows drawing lessons from deficiencies and failures of the aviation system.

In this context, reports are submitted confidentially and front-line personnel should not be subject to any prejudice on the basis of the information provided, except in cases of 'willful misconduct' or "where there has been manifest, severe and serious disregard with respect to an obvious risk and profound failure of professional responsibility to take such care as is evidently required in the circumstances" causing foreseeable damage to a person or to property, or seriously compromising the level of aviation safety (*i.e.* in case of 'gross negligence') (recital 37).

In order to ensure appropriate confidentiality of front-line operators regarding the occurrence reporting system, the aforementioned recital 34 points out that the information contained in occurrence reports should be protected appropriately and should not be used for purposes other than maintaining or improving aviation safety, in respect of the 'just culture' rules. In this context, limitations on the transmission of reporters' personal details should also be ensured.

In order to increase the confidence of individuals, the handling of occurrence reports would ensure protection of the reporter's identity in order to promote a 'just culture' (recital 40 of Regulation No. 376/2014). This recital is closely linked to the recital 45 which, in view of promoting a 'just culture', highlights that "The handling of the reports shall be done with a view to preventing the use of information for purposes other than safety, and shall appropriately safeguard the confidentiality of the identity of the reporter and of the persons mentioned in occurrence reports".

Recital 44, on the use of occurrence reports, emphasises that

Nevertheless, in the context of developing a 'just culture' environment, Member States should retain the option of extending the prohibition on using occurrence reports as evidence against reporters in administrative and disciplinary proceedings to civil or criminal proceedings.

That requires a commitment by all Member States to prohibit the use of the occurrence reports as evidence in administrative, disciplinary, civil or criminal proceedings.

Recital 38 extends the subjective scope of protection, reserved for the reporters, to people mentioned in the occurrence reports ("in order to encourage reporting of occurrences, it should be appropriate to protect not only reporters, but also persons mentioned in the occurrence reports concerned").

However, such protection should not exonerate those persons from their obligations under this Regulation.

It means that where a person has an obligation to report an occurrence, and intentionally fails to do it, that person should lose his protection and should be subject to penalties in application of the present Regulation. In other words, 'just culture' does not cover the omission of a reporting duty.

Other explicit references to 'just culture' principles are contained in several points of Article 16. In particular, paragraph 6 states that, without prejudice to applicable national criminal law, Member States shall not institute proceedings in respect to "unpremeditated or inadvertent infringements of the law" which are taken into consideration only because they have been reported. This means that this provision shall not apply to premeditated and conscious violations of the law that involve criminal proceedings.

In fact, paragraph 6 of this Article—"without prejudice to applicable national criminal law"—states that "Member States shall refrain from instituting proceedings of unpremeditated or inadvertent infringements of the law which come to their attention only because they have been reported pursuant to Articles 4 and 5".

This provision, which reproduces almost in full the text of the now abolished Article 8(3) of the aforementioned Directive 2003/42/EC, does not allow for any prosecution within the Union against occurrence reporters for their actions, omissions or decisions, in the case of 'unpremeditated' or 'inadvertent' infringement of the law. As already mentioned, 'unpremeditated' violation refers to unintentional conduct, which is not characterized by willful intent and is not planned in advance, whereas 'inadvertent' refers to an unconscious infringement. Both types of behaviours have to be tolerated.

In case of disciplinary or administrative proceedings, information contained in occurrence reports shall not be used against the reporters, nor the persons mentioned in occurrence reports (Article 16(7)).

The same Article 16(9) stresses that employees and aviation personnel who report or are mentioned in occurrence reports shall not be subject to any prejudice by their employer or by the organization they work for because of the information provided by the reporter.

Therefore, Article 16(10) clarifies that the previous provision shall not apply in cases of 'wilful misconduct' or "where there has been a manifest, severe and serious disregard of an obvious risk and profound failure of professional responsibility to take such care as is evidently required in the circumstances, causing foreseeable damage to a person or property, or which seriously compromises the level of aviation safety".

Obviously, the phrase marked included in quotation marks refers to the concept of 'gross negligence'.

More explicitly, the same Article 16(11), states: "Each organisation established in a Member State shall, after consulting its staff representatives, adopt internal rules describing how 'just culture' principles, in particular the principle referred to in paragraph 9, are guaranteed and implemented within that organisation".

This paragraph introduces a clear obligation for aviation organisations to adopt internal rules aspiring to a 'just culture' and, in particular, to the principle that excludes any prejudice for the reporter as a consequence of the reporting, with commitment of each Member State to explain how it intends to apply such principles.

In the light of the above, Article 6 requires that for each organisation established in the EU territory, each Member State and EASA must designate one or more persons or authorities to handle the collection, evaluation, processing, analysis and storage of details of occurrences reported. The aim of this provision is to avoid the use of information on aviation occurrences for purposes other than prevention (for example, to initiate criminal or disciplinary proceedings)[75] and at the same time protect the confidentiality of the reporter's identity, with a view to promoting a 'just culture' (Article 6(2)).

In conclusion, in the face of the contents of Regulation No. 376/2014, it can be concluded that these principles have fully inspired the new European legislation.

In fact, European lawmakers revel a clear intention to adopt the distinction between acceptable (honest mistake) and unacceptable ('wilful misconduct' or 'gross negligence') behaviour, representing the focal point of a 'just culture', applicable to administrative, disciplinary and judicial proceeding,[76] indistinctly.

Therefore, although the 'blame culture' is antithetical to the 'safety culture', the first one continues to exist even in organizations that have formally implemented a SMS.[77]

Consequently, the question is "Does the organization have the elements of a SMS not only documented but also in place and implemented in actual operations?"[78]

In order to meet this objective, an atmosphere to ensure that mistakes are not hidden, but "systematically reported, analysed and used as an information source for learning, mitigating measures and safety improvements"[79] is needed. It is an environment that reflects the real commitment to civil aviation safety at all levels.

Moreover, another issue deserves consideration.

Actually, despite the introduction of a legal definition of 'just culture', the European legislature did not introduce new safety regulatory requirements for the implementation of this culture, but it only confined itself to monitor[80] the implementation.

It should be stressed that the application of 'just culture' principles requires a corresponding national legal framework because the administration of justice is the responsibility of the Member States and their national authorities.

In fact, particularly in the criminal law field, the administration of justice constitutes one of the pillars of the sovereign functions of the State.

In international law, the 1944 Chicago Convention, mentioned above, explicitly confirms that every State has "complete and exclusive sovereignty over the airspace above its territory" (Article 1)[81] and that certainly includes the administration of justice.

---

[75]So Taviano (2018), p. 253 f.

[76]Dekker and Breakey (2016), p. 187 ff.

[77]Ulfvengren and Corrigan (2015), p. 219 ff.

[78]So Koivu (2013), p. 66.

[79]*Ibid.*

[80]Petschonek (2011); Marchetti (2017).

[81]Diederiks-Verschoor and Butler (2006), p. 17 f.

Therefore, although the EU Member States chose to delegate or pool certain functions at European level, the sanction system and criminal jurisdiction generally remain embedded at national level with only a few exceptions.

2.1.3   At National Level (Italian Legal System)

The enactment of 'just culture' principles into the Italian legal system was the result of the application of Community law.

Although until the 1990s we cannot find express references to the 'just culture' in the Italian laws, certain provisions deserve to be highlighted because they contain the first signs of the development of this culture.

First of all, Legislative Decree No. 66 of 25 February 1999,[82] which implemented Directive 94/56/CE into Italian law, needs be discussed here.

Article 12 of this Decree, laying down provisions concerning accident and incident reports, stresses that the report shall state that the sole objective of the investigation is the prevention of future accidents and incidents and shall protect the anonymity of the persons involved in the single occurrence.

Subsequently, Article 8 of the Italian Legislative Decree No. 213 of 2 May 2006[83] implemented Directive 2003/42/EC[84] on occurrence reporting in civil aviation. It laid down provisions on the protection of information. In particular, it stated that occurrences reported must be entered into the ENAC database in such a way that does not allow reporters' identification. The Italian National Agency for the Safety of Flight (ANSV)[85] establishes the procedures to guarantee the immediate cancellation of personal data from the reports. Employees who report incidents of which they may have knowledge are not subjected to any prejudice by their employer.

In order to complete the legal framework, we can also refer to the ENAC Circular of 2011[86] on the mandatory reporting of accidents, serious incidents and aviation occurrences where it states:

---

[82]Legislative Decree No. 66 of 25 February 1999 "Establishment of the National Agency for the Safety of Flight and amendments to the navigation code, pursuant to Council Directive 94/56/EC of 21 November 1994" (G.U. No. 67 of 22 March 1999). For a commentary Franchi (2000), p. 595 ff.; Pellegrino (2007), p. 333 ff.

[83]Legislative Decree No. 213 of 2 May 2006 "Implementation of Directive 2003/42/EC on occurrence reporting in civil aviation" (G.U. No. 137 of 15 June 2006). For a commentary, see Franchi (2006), p. 137 ff.; Pellegrino (2007), p. 390 ff.

[84]Repealed by Article 23 of Regulation (EU) No. 376/2014.

[85]Authority created by the Italian Legislative Decree No. 66/99, mentioned above. For a commentary see Franchi (2000), p. 595 ff.

[86]ENAC GEN-01B Circular on Mandatory Occurrence Reporting System eE-MOR, General Series, of 11 April 2011, paragraph 6, in https://www.enac.gov.it/ContentManagement/informa tion/P2004690470/GEN_01B_accessible.pdf.

the reporting system should not be based on punitive principles, but must be aimed primarily to identify eventual organizational deficiencies in order to optimise the internal process, to the benefit of safety [...] However, this provisions do not excuse grossly negligent behaviours that should be isolated and neutralized by the organization for a global approach to safety (para. 6).

As we have seen, the first legislation that contains an express reference to 'just culture' and its definition is the Commission Regulation (EU) 691/2010, followed by Regulation (EU) No. 376/2014.

Furthermore, a soft law document, a recent 'Technical & Professional Manual' (TPM)[87] by IFATCA (International Federation of Air Traffic Controllers' Associations)[88] contains the same definition of 'just culture' as mentioned above. In fact, this Manual, reiterating the common definition, considers a 'just culture' as 'a culture in which front line operators or others are not punished for actions, omissions or decisions taken by negligence, willful violations and destructive acts are not tolerated'.

Finally, the recent document entitled 'State Safety Programme – Italy',[89] elaborated in 2017 by the Italian Civil Aviation Authority (ENAC),[90] also contains the same agreed definition of 'just culture'[91] and requires this authority to bring its 'safety policy'[92] in line with the 'just culture' principles.

2.2 Considerations on the 'Just Culture' Common Definition

In the light of the common definition of 'just culture', contained in the international and European legislation and regulation, and on the basis of the previous interpretation of this concept, some issues need be taken into consideration.

---

[87]See IFATCA (International Federation of Air Traffic Controller's Association), Technical & Professional Manual, The permanent record of the Federation's Technical & Professional Policies (2016), Montreal (Canada), LM 11.2.1 (Just Culture, Trust and Mutual Respect), p. 4248.

[88]IFATCA unites the professional associations of air traffic controllers from around the world. See European Union, European Flight and Aviation Safety Regulations Handbook, System, procedures and Important Regulations, vol. 1, International Business Publications, Washington, USA, 2009, pp. 84 f.

[89]ENAC, State Safety Programme – Italy, edition 3, February 2017.

[90]Established on 25 July 1997 by Legislative Decree No. 250 of 25 July 1997 (G.U. No. 177 of 31 July 1997) as the National Authority committed to oversee the technical regulation, the surveillance and the control in the civil aviation field.

[91]For the definition of 'just culture' ("a culture in which front line operators or others are not punished for actions, omissions or decisions taken by negligence, willful violations and destructive acts are not tolerated"), see ENAC, State Safety Programme – Italy, edition 3, February 2017, p. 7.

[92]See the definition of 'safety policy' in ENAC, State Safety Programme – Italy, edition 3, February 2017, p. 8 and lett. j), p. 9.

This definition fleshes out in legal form the concepts originally developed by Reason, who defined 'just culture' as "an atmosphere of trust in which people are encouraged (even rewarded) for providing essential safety-related information but in which they are also clear about where the line must be drawn between acceptable and unacceptable behaviour".

This definition is focused on the dividing line between 'gross negligence', 'willful violations' and 'destructive acts', on the one hand, and actions, omissions or decisions by front-line operators, consistent with experience and training, on the other hand.

Regarding this, the first question is: what are front-line operators ('flo')?

In the aviation industry, they are workers[93] who are in direct contact with hazard. In particular, the following personnel falls within the category of front-line operators: air traffic controllers,[94] pilots, flight crew, cabin crew, maintenance personnel,[95] airport managers and all other people who provide important information about aviation safety problems.

All behaviours understood as 'mere negligence'[96] or 'honest error' of front-line operators should be tolerated, while 'gross negligence', 'willful misconduct' and, in particular, dangerous and destructive acts are not acceptable and thus are punishable.

This means that a totally 'no blame' culture is "neither feasible nor desirable"[97]: some level of accountability is needed when a mishap[98] occurs. Full amnesty for all unlawful and unsafe acts would lack credibility in the public opinion.[99]

On the contrary, a 'honest mistake' should not be punished in order to encourage reporting of failures and information sharing from the personnel involved.

In this context, it was pointed out that for the improvement of aviation safety a 'feedback knowledge derived from a systematic accident/incident data collection and analysis'[100] is needed, but it is hampered by a "growing fear of litigations and threat of sanctions against individuals, particularly if they were partly or fully responsible for an accident they are involved in or an incident they reported".[101]

---

[93]Furnham (1997).

[94]Ruitenberg (2002), p. 22 ff.; Rizzo (2009); Dekker (2010b), p. 31 ff.

[95]Helmereich and Clyton Foushee (1993), p. 3 ff.; Hobbs (1997); Gaur (2005), p. 503 ff; Harris (2011). See also Fanara (2000); La Torre et al. (2006); Busti et al. (2017).

[96]Dovere (2016), p. 1023 ff.

[97]Reason (1997), p. 195.

[98]Gibb et al. (2016), p. 4.

[99]See Dekker (2010a), p. 275 ff.

[100]See Working Paper of ICAO Assembly, 36th Session, Technical Commission, A36-WP/232, TE/76, 18 September 2007, para. 1.1, 2.

[101]See Working Paper of ICAO Assembly, 36th Session, Technical Commission, A36-WP/232, cited above, 1.2, p. 2.

References

Antonini A, Franchi B (eds) (2005) Diritto aeronautico a cent'anni dal primo volo. Giuffré, Milano

Baumgartner M, Schorer R (eds) (2017) Just Culture Manual for ATCO, ANSE & ATSEP. Behavior after an incident and further proceedings, SwissATCA

Bernabei V, Barbafina F (2015) Il Safety Management System nel campo dell'aviazione. In: Pellegrino F (ed) Legislation and regulation of risk management in aviation activity, vol II. Giuffré, Milano, pp 163–170

Bieder C, Bourrier M (eds) (2017) Trapping safety into rules: how desirable or avoidable is proceduralization? Taylor & Francis Group, Boca Raton (Florida)

Brooker P (2007) The European single sky needs high quality, simple incident reporting. Air Traffic Technology International, p 3 ff

Bufo M (2012) Le competenze degli ANSPs. Il servizio informazioni volo (FIS) e il servizio informazioni volo aeroportuale (AFIS). In: Pellegrino F (ed) Air navigation rules and practices in Europe: towards harmonization. Giuffré, Milano, pp 163–190

Busti S, Signorini E, Simoncini GR (eds) (2017) L'impresa aeroportuale a dieci anni dalla riforma del codice della navigazione: stato dell'arte. Giappichelli, Torino

Calleja Crespo D, Mendes de Leon P (eds) (2011) Achieving the Single European Sky: goals and challenges. Kluwer, Alphen aan den Rijn

Cassatella A (2017) Appeals before the European Aviation Safety Agency. In: Marchetti B (ed) Administrative remedies in the European Union: the emergence of a quasi-judicial administration. Giappichelli, Torino, pp 21–54

Castellani G (2015) Responsabilità sociale di impresa. Ragioni, azioni e reporting, Maggioli, Sant'Arcangelo di Romagna (Rimini)

Cohen-Charash Y, Spector PE (2001) The role of justice in organizations: a meta-analysis. Organ Behav Hum Decis Process 86(2):278–321

Coman-Kund F (2018) European Union agencies as global actors: a legal study of the European Aviation Safety Agency, Frontex and Europol. Routledge, Taylor & Francis, Abingdon, Oxon, New York

Cook A (ed) (2016) European Air Traffic Management. Principles, practice and research (safety, transparency and 'just culture'). Ashgate, Aldershot (Hampshire)

Corrigan S (2003) Comparative analysis of safety management systems and safety culture in aircraft maintenance. Trinity College, Dublin

De Franchis F (1985) Law dictionary. English-Italian, Giuffrè, Milano, vol 1, p 1402

Dekker S (2010a) Balancing 'no blame' with accountability. N Engl J Med 362(3):275–276

Dekker S (2010b) Pilots, controllers and mechanics on trial: cases, concerns and countermeasures. Int J Appl Aviat Stud (IJAAS) 10(1):31–49

Dekker S (2012) Just culture. Balancing safety and accountability. Ashgate, Aldershot

Dekker SW, Breakey H (2016) 'Just culture': improving safety by achieving substantive, procedural and restorative justice. Safety Sci 85:187–193

Di Carlo R (2015) Dalla compliance alla performance. Possibili implicazioni dell'evoluzione delle norme di Safety Risk Manafùgement nel settore dell'aviazione civile. In: Pellegrino F (ed) Legislation and regulation of risk management in aviation activity, vol II. Giuffrè, Milano, pp 203–227

Di Renzo (2015) Legislation and regulation of risk management in aviation activity. Civil Aviation Safety Management System. In: Pellegrino F (ed) Legislation and regulation of risk management in aviation activity, vol II. Giuffré, Milano, pp 155–162

Diederiks-Verschoor IHP, Butler MA (2006) An introduction to air law. Kluwer Law International, The Hague

Dovere S (2016) Prospettive della responsabilità penale colposa nel settore aeronautico (l'espressione 'ultra limes' della colpa lieve). Responsabilità civile e previdenza 3:1023–1038

Espinola S, Costa M, Maurino D (2005) Guidance material addresses concerns about protection of safety information. ICAO J 61(6):26–28

Fanara E (ed) (2000) La nuova disciplina del trasporto aereo. Messina

Franchi B (2000) L'Agenzia Nazionale per la sicurezza del volo. In: Fanara E (ed) La nuova disciplina del trasporto aereo. Messina, p 595 ff

Franchi B (2006) La sicurezza del passeggero nel trasporto aereo. In: Masala L, Rosafio E (eds) Trasporto aereo e sicurezza del passeggero. Giuffrè, Milano, p 145 ff

Franchi B (2012) Le autorità investigative per la sicurezza dell'aviazione civile dopo l'entrata in vigore del Regolamento U.E. 996/2010. In: Pellegrino F (ed) Air navigation rules and practices in Europe: towards harmonization. Giuffré, Milano, pp 41–56

Furnham A (1997) The psychology of behaviour at work. Psychology Press, Hove England

Gaur D (2005) Human factor analysis and classification system applied to civil aircraft accidents in India. Aviat Space Environ Med 76:501–505

Giannini A (1952) L'organizzazione dell'ICAO. Rivista aeronautica, pp 579–584

Gibb R, Gray R, Scharff L (2016) Aviation visual perception: research, misperception and mishaps. Taylor & Francis Group, Routdlege, London, New York

Giemulla EM, Weber L (2011) International and EU aviation law: selected issues. Kluwer, Dordrecht

Guarrera G (2015) Safety Airport Management. In: Pellegrino F (ed) Legislation and regulation of risk management in aviation activity, vol II. Giuffré, Milano, pp 171–178

Harris D (2011) Human performance on the flight deck. Ashgate, Furnham

Harris D (ed) (2013) Engineering psychology and cognitive ergonomics. Applications and services. Springer-Verlag, Berlin, Heidelberg

Helmereich RL, Clyton Foushee H (1993) Why CRM? Empirical and theoretical bases of human factors training in aviation. In: Weiner EL, Kanki BG, Helmereich RL (eds) Cockpit resource management. Academic Press, San Diego, pp 3–57

Hobbs A (1997) Human factor in airline maintenance: a study of incident reports. Bureau of Air Safety investigation, Canberra (Australia)

Hudson PTW (2002) Real time decision making in sport. In: Moorman PP, Pipers R (eds) Proceeding Congress Sportpsychologie. Leiden University, Leiden, p 74 ff

Hudson PTW (2009) Process indicators: managing safety by the numbers. Saf Sci 47:483–485

Koivu H (2013) Just and safety: the art of making mistakes. People sell washing machines – robots fly aeroplanes? HindSight 18:66 ff

La Torre U, Moschella G, Pellegrino F, Rizzo MP, Vermiglio G (eds) (2006) Studi in memoria di Elio Fanara, vol 1. Giuffrè, Milano

Lerner JS, Tetlock PE (1999) Accounting for the effects of accountability. Psychol Bull 125 (2):255–275

Licu T, van Dam R (2013) Just culture in aviation: dynamics and deliverables. HindSight 18:18–21

Licu T, van Dam R (2014) Just culture in aviation: dynamics and deliverables. In: Pellegrino F (ed) Legislation and regulation of risk management in aviation activity, vol I. Giuffré, Milano, pp 55–65

Licu T, Baumgartner M, van Dam R (2013) Everything you always wanted to know about just culture (but we afraid to ask). HindSight 18:14–17

Ludwig DA (2009) Safety management systems for airports, vol 2. Guidebook, Transportation Research Board, Washington

Marchetti B (ed) (2017) Administrative remedies in the European Union: the emergence of a quasi-judicial administration. Giappichelli, Torino

Marino A (2013) Agenzie e Autorità di regolazione del trasporto nel diritto comunitario e interno. Jovene, Napoli

Masala L, Rosafio E (eds) (2006) Trasporto aereo e sicurezza del passeggero. Giuffrè, Milano

McDonald N, Corrigan S, Daly C, Cromie S (2000) Safety management systems and safety culture in aircraft maintenance organisations. Saf Sci 34:151–176

Michaelides-Mateou S, Mateou A (2014) Notification and reporting of aircraft accidents or incidents. In: Pellegrino F (ed) Legislation and regulation of risk management in aviation activity, vol I. Giuffré, Milano, p 265 ff

Monaco R (1981) Le funzioni dell'OACI. In: Monaco R (ed) Scritti di diritto delle organizzazioni internazionali. Giuffrè, Milano, p 403 ff

Moorman PP, Pipers R (eds) (2002) Proceeding Congress Sportpsychologie. Leiden University, Leiden

Muchinsky PM (1997) Psychology applied to work, 5th edn. Brooks/Cole, Pacific Grove (California), United States

Murray SR (1997) Deliberate decision making by aircraft pilots: a simple reminder to avoid decision making under panic. Int J Aviat Psychol 7:83–100

Nastro V (1993) Assistenza al volo e controllo del traffico aereo. Hoepli, Milano

O'Leary M, Chappel SL (1997) Confidential incident reporting systems create vital awareness of safety problems. ICAO J 51:11–13

Orasanu JM (1993) Decision-making in the cockpit. In: Weiner EL, Kanki BG, Helmereich RL (eds) Cockpit resource management. Academic Press, San Diego, pp 137–172

Pellegrino F (2007) Sicurezza e prevenzione degli incidenti aeronautici. Giuffré, Milano

Pellegrino F (2009) La Direttiva 2006/23/CE in materia di licenza comunitaria dei controllori di volo. In: Rizzo MP (ed) La gestione del traffico aereo: profili di diritto internazionale, comunitario ed interno, vol 242. Giuffrè, Milano, pp 211–232

Pellegrino F (ed) (2012) Air navigation rules and practices in Europe: towards harmonization. Giuffré, Milano

Pellegrino F (ed) (2014) Legislation and regulation of risk management in aviation activity, vol I. Giuffré, Milano

Petrick-Felber N (2014) Liberalizing Europe's Skies – a failure? An analysis of airline entry and exit in the post-liberalized German Airline Market, 1993–2006. Anchor Academic Publisher, Hamburg

Petschonek SL (2011) Developing the just culture assessment tool: a method for measuring individual cultural perceptions in a healthcare setting. University of Memphis, Memphis

Pifisterer H (2017) European regulation of aerodrome safety management system in the EASA system. Kassel University

Preti F (2012) Il ruolo del Performance Review Body nel Cielo Unico Europeo: la regolazione delle prestazioni nell'Unione europea. In: Pellegrino F (ed) Air navigation rules and practices in Europe: towards harmonization. Giuffré, Milano, pp 113–119

Randazzo V (2004) Alcuni profili problematici relativi all'attribuzione di funzioni all'Agenzia europea per la sicurezza aerea. Diritto dell'Unione Europea 4:847–867

Reason J (1997) Managing the risks of organizational accidents. Ashgate, Aldershot, e-book published in 2016

Reiman T, Pietikäinen E (2010) Indicators of safety culture – selection and utilization of leading safety performance indicators. Report number: 2010:07

Rizzo MP (2006) Il pacchetto di regolamenti comunitari per la realizzazione del "cielo unico europeo". In: La Torre U, Moschella G, Pellegrino F, Rizzo MP, Vermiglio G (eds) Studi in memoria di Elio Fanara, vol 1. Giuffrè, Milano, pp 407–441

Rizzo MP (ed) (2009) La gestione del traffico aereo: profili di diritto internazionale, comunitario ed interno. Giuffrè, Milano

Rossi dal Pozzo F (2014) EU legal framework for safeguarding air passenger rights. Springer, Heidelberg, New York, Dordrecht, London

Roughton J, Mercurio J (2002) Developing an effective safety culture: a leadership approach. Butterworth-Heinemann, Woburn, Massachusetts

Ruitenberg B (2002) Court case against Dutch controllers. The Controller, p 22 ff

Sardella B (1995) "Eurocontrol" non è un'impresa secondo le norme del diritto comunitario. Giustizia civile, p 12 ff

Schwenk W, Schwenk R (1998) Aspects of international cooperation in air traffic management. Martinus Nijhoff Publishers, The Hague, Boston, London, p 129 ff

Sciolla Lagrange A (1990) Organizzazione dell'aviazione civile internazionale (OACI). Enciclopedia Giuridica Treccani, XXXI, p 1 ff

Sharpe VA (ed) (2015) Accountability: patient safety and policy reform. Georgetown University Press, Washington

Spadoni AS (2001) Eurocontrol e la funzione unificatrice dello spazio aereo europeo: rafforzamento o svuotamento del principio della sovranità degli Stati. Rivista di diritto pubblico e scienze politiche 2:259–267

Stadler G (2009) The role of eurocontrol in the implementation of the Single European Sky. In: Rizzo MP (ed) La gestione del traffico aereo: profili di diritto internazionale, comunitario ed interno. Giuffrè, Milan, pp 267–279

Starrantino C, Finocchiaro M (2014) Just culture. A "new" approach to safety. In: Pellegrino F (ed) Legislation and regulation of risk management in aviation activity, vol I, pp 23–54

Stolzer AJ, Goglia JJ (eds) (2015) Safety management systems in aviation, 2nd edn. Routledge, Taylor & Francis Group, Abingdon, Oxson

Stolzer AJ, Halford CD, Goglia JJ (2011) Implementing safety management systems in aviation. Ashgate, Farhnam (England)

Taviano M (2018) Il quadro giuridico della just culture nel settore aeronautico italiano e i passi indietro nel d.lgs. n. 173/2017. Rivista di diritto dell'economia, dei trasporti e dell'ambiente, Giureta XVI:245–267

Togan S (2016) The liberalization of transportation services in the EU and Turkey. Oxford University Press, p 224 ff

Trovò L (2008) Eurocontrol: in assenza di attività economica non scattano le norme sulla concorrenza. Diritto dei trasporti:163–170

Trovò L (2011) Il processo d'integrazione degli spazi aerei europei: dalla riorganizzazione in blocchi funzionali verso la globalizzazione dell'Air Traffic Management (ATM). Rivista di diritto dell'economia, dei trasporti e dell'ambiente, Giureta:24 ff

Turco Bulgherini E (2009) Le compentenze dell'ENAV in materia di controllo del traffico aereo nel quadro della revisione della parte aeronautica del codice della navigazione. In: Rizzo MP (ed) La gestione del traffico aereo: profili di diritto internazionale, comunitario ed interno, pp 331–372

Tytgat L (2012) The Single European Sky Regulation in Europe: new scenarios. In: Pellegrino F (ed) Air navigation rules and practices in Europe: towards harmonization. Giuffré, Milano, pp 93–96

Ulfvengren P, Corrigan S (2015) Development and implementation of a safety management system in a Lean Airline. Cogn Technol Work 17(2):219–236

Valente P (2005) Single European Sky: cielo unico europeo. In: Antonini A, Franchi B (eds) Diritto aeronautico a cent'anni dal primo volo. Giuffré, Milano, pp 203–206

Van Antwerpen N (2009) Cross-border provision of air navigation services with specific reference to Europe. Safeguarding transparent lines of responsability and liability. Kluwer, Alphen aan den Rijn

Vaughan D (1999) The dark side of organizations: mistake, misconduct, and disaster. Annu Rev Social 25:271–305

Vernizzi S (2017) Gestore aeroportuale e bird strike, in particolare, il caso Antonov: tutti colpevoli. No, anzi, tutti assolti! In: Busti S, Signorini E, Simoncini GR (eds) L'impresa aeroportuale a dieci anni dalla riforma del codice della navigazione: stato dell'arte. Giappichelli, Torino, pp 255–268

Vincenzi L (2010) L'Unione europea rafforza il cielo unico. Diritto marittimo 1–2:316–319

Weiner EL, Kanki BG, Helmereich RL (eds) (1993) Cockpit resource management. Academic Press, San Diego

Xerri A (2012) Cielo unico europeo: riflessioni su un diritto aeronautico europeo. In: Pellegrino F (ed) Air navigation rules and practices in Europe: towards harmonization. Giuffrè, Milano, p 67 ff

Chapter 3
Legitimate and Illegitimate Behaviour in Aviation

3.1 The Fine Line Between Acceptable and Unacceptable Behavior

In the light of the analysis conducted hitherto, in my opinion, only a few unsafe human actions are deliberate and require criminal-law punishments.

'Gross negligence', 'willful violations' or destructive acts cannot be considered as honest mistakes: therefore, they are not tolerated and are not exempt from prosecution.

On the contrary, we know we can learn a lot from the so-called 'honest mistakes' in aviation.

In order to give effective force to 'just culture' principles, a constructive environment should be promoted, in which people are encouraged to provide essential safety-related information, but in which they are also clear about where the line between acceptable and unacceptable behaviour must be drawn.

In this regard, the above mentioned Eurocontrol Policy Model recognises that, even though nothing should prevent criminal prosecutions in the event of intentional wrongdoing or 'gross negligence', no prosecution should be brought against individuals for actions, omissions or decisions that reflect the conduct of a reasonable person even when they may have caused an unpremeditated or inadvertent infringement of the applicable law.

To this purpose, the Eurocontrol's ESARR Advisory Guidance entitled "Establishment of 'Just Culture' Principles in ATM Safety Data Reporting and Assessment"[1] stresses that "the difficult task is to discriminate between the truly 'bad

---

[1]Eurocontrol Safety Regulation Commission, ESARR Advisory Material/Guidance Document (EAM 2/GUI 6) "Establishment of 'Just Culture' principles in ATM Safety Data Reporting and Assessment", ESARR Advisory Material, edition 1, 31 March 2006, 17. See also Eurocontrol's Performance Review Unit (PRU), Legal and Cultural Issues in relation to ATM. Safety Occurrence

© Springer Nature Switzerland AG 2019
F. Pellegrino, *The Just Culture Principles in Aviation Law*, Legal Studies in International, European and Comparative Criminal Law 3,
https://doi.org/10.1007/978-3-030-23178-1_3

behaviours' and the unsafe acts for which discipline is neither appropriate nor useful. It is necessary to agree on a set of principles for drawing this line".

All definitions of a 'just culture' and all references to it, contained in the EU legislation and both in legal or technical documents, draw a line between acceptable and unacceptable behavior (unpunishable and punishable behavior). Yet this line is also exactly what makes it difficult to implement this culture in practice, and to separate it from the field of criminal law.

Although the demand of a 'just culture' definition has been met, there is still a need for uniform and agreed-upon definitions about what is acceptable behaviour, and what is not acceptable,[2] e.g. about what is willful misconduct or gross negligence and what is an 'honest mistake'. Even today, these concepts are subject to a case-by-case evaluation by the competent judicial authorities.

In this context, many questions arise: who gets to draw the dividing line?[3] How to identify the dividing line? Who will determine whether a mistake was made by an operator acting in a responsible and reasonable manner or whether this was a case of 'gross negligence', 'wilful misconduct' or criminal intention?

Such a check can only be made by the prosecutors and ultimately by the courts in criminal proceedings. Nevertheless, prosecutors and judges usually do not have the technical skills to reenact what happened. In most cases, they lack the appropriate knowledge and practical experience to appreciate an aviation occurrence from the perspective of reasonable aviation professionals. Therefore, to reach conclusions about the dynamics of an occurrence, they rely on credible experts, usually recruited from the aviation community, who have the skills and the ability to objectively evaluate the behaviour of aviation operators. The aviation community representatives often provide many of the elements that will support and justify a subsequent prosecution.

However, the goal of a 'just culture' approach is not to transfer the task of evaluating these legal aspects to the aviation community representatives, who have exclusively technical competences.

In some jurisdictions, safety investigation reports, written from an operational and technical perspective ('without apportioning blame or liability'),[4] are admissible as evidence in criminal proceedings. In other countries, such as in Italy, these reports are included in the trial file and can be evaluated by the judge on a discretionary basis.

This means that legal conclusions about the guilt should be drawn by the judges based on findings contained in the report prepared by appointed experts. This is done by applying principles and rules of law to qualify the operator's behavior as acceptable or criminal. To this end, they should consider the front-line behavior

---

Reporting in Europe. Outcome of a Survey conducted by the Performance Review Unit in 2005–2006, Bruxelles, September 2006.

[2]In particular, see Baumgartner (2011), p. 299 ff.; Baumgartner et al. (2011), p. 281 ff.

[3]Dekker (2016), para. 1.

[4]So Article 16 of Reg. (EU) No. 996/2010, cited above.

with flexibility, given the specific characteristics of the aviation profession and considering the difficult situations which can occur during flight activity. These particularities should require a broad interpretation of the concepts of negligence, imprudence and malpractice by the courts, at least in the aviation sector.

3.2   Current Definitions of 'Wilful Misconduct' and 'Gross Negligence'

'Willful misconduct' and 'gross negligence' are concepts that are currently defined in no international or EU legislation because of the significant differences existing between common and civil law systems.

In fact, the difficulty to find such definitions depends on the differences existing between countries on the concepts of 'willful misconduct' and 'gross negligence'.

As has been observed, the behaviours that the courts consider as falling under those concepts are investigated on a case-by-case basis.

Actually, it is difficult to differentiate an act of 'gross negligence' from one of 'willful misconduct' because there is a grey area between these two extremes: it leaves the possibility of different judicial interpretations in the different countries.

With respect to the application of the concepts of 'willful misconduct' and 'gross negligence' in civil law systems, one must recall the principle *culpa lata dolo aequiparatur* of the classical Roman law. In this context, *culpa lata* broadly corresponds to 'gross negligence' and *dolus* broadly corresponds to 'willful misconduct'. What is the reason behind such an equivalence? The 'willful misconduct' standard is similar to the 'gross negligence' standard.

In recent years, contractual definitions for these terms have been developed in special sectors, for example in the oil and gas industry.

In the standard form "Construction, ownership and operation agreement" elaborated by the Petroleum Joint Venture Association (PJVA),[5] a more flexible definition of 'gross negligence' has been included in Article 1 as follows:

> Gross Negligence means: (i) a marked and flagrant departure from the standard of conduct of a reasonable person acting in the circumstances at the time of the alleged misconduct, or (ii) such wanton and reckless conduct or omissions as constitutes in effect an utter disregard for harmful, foreseeable and avoidable consequences.[6]

---

[5]This Association represents and supports individuals and organizations involved in petroleum joint venture. PJVA provides a multi-disciplinary forum to address current petroleum joint venture issues. Its mission is to enhance the efficiency and effectiveness of petroleum joint venture relationships and activities. See Duval et al. (2009), pp. 286 ff.

[6]PJVA Operating Agreement, October 2003, Article 1, s 101 (f) Definitions.

The "Operating Procedure" developed by the Canadian Association of Petroleum Landmen (CAPL),[7] operating in the same sector, provided the following definition:

> Gross Negligence or Wilful Misconduct means any act or failure to act (whether sole, joint or concurrent) by a person that was intended to cause or was in reckless disregard of, or wanton indifference to, the harmful consequences to the safety or property of another person which the person acting or failing to act knew, or should have known, would result from such act or omission, provided that Gross Negligence or Wilful Misconduct does not include any act or failure to act insofar as it: (i) constituted mere ordinary negligence; or (ii) was done or omitted in accordance with the express instructions or approval of all parties.[8]

Both the PJVA and CAPL definitions seemed to equate 'gross negligence' and 'willful misconduct', with regard to the effects. In fact, the PJVA definition also provides for liability due to 'gross negligence', in circumstances that do not involve the intention to harm. The CAPL definition provides for liability for "willful liability" but it is not clear if it is also extended to "gross negligence". In fact, according to the last definition, there is no liability unless there was a conduct that is intentional, reckless or wanton in nature.

But more recently, the CAPL definition has been modified with the inclusion of a new item (i) as follows: "a marked and flagrant departure from the standard of conduct of a reasonable operator acting in the circumstances at the time of the alleged misconduct".[9]

This modification has been justified by the following reasons: "the 2007 definition was inadvertently written in a way that focused on the willful or wanton misconduct, reckless disregard components, without sufficient reference to the gross negligent component".[10]

The intent to align the Canadian text more closely to the PJVA definition is clear.

Although a number of legal systems equate, 'gross negligence' and intentional harm/wrongdoing from a conceptual view, the following question arises: What is the legal difference between 'gross negligence' and 'wilful misconduct', given that the two expressions are often considered equivalent with respect to legal consequences in civil-law instruments?

This link between 'wilful misconduct' and 'gross negligence', based on the legal consequences of the alleged conduct, can be easily accepted in civil proceedings, where the conceptual distinction is almost irrelevant, but not in criminal proceedings, considering the real differences between these expressions.

For our purposes it is necessary to clearly differentiate them.

Actually, it is impossible to provide a wide definition that represents a certain parameter to assess the various types of conduct.

---

[7]2007 CAPL Operating Procedure, update in 2015. See Duguid (2006); Grosse (2010); Rodhes (2013); Sweeney (2016), footnotes pp. 666 ff.

[8]2007 CAPL Operating Procedure, (Definitions).

[9]2015 CAPL Operating Procedure Annotated (Part I), p. 4.

[10]See 2015 CAPL Operating Procedure, Detailed Table Outlining Modifications to the 2007 CAPL Operating Procedure, p. 3.

Within the meaning described in most of the legal dictionaries and case law, 'gross negligence' is a failure to use such a degree of care, precaution or vigilance (*i.e.* serious carelessness) that any reasonable and prudent person would use under the same or similar circumstances and normally refers to voluntary risk-taking.[11] In other words, the person accused of 'gross negligence' consciously takes an unjustifiable risk that a 'reasonable person' would not take. It is therefore a serious deviation from a reasonable and prudent person's standard, characterized by high levels of negligence (gravity). This occurs when a person, who foresees a negative result of his action, acts recklessly and consciously or voluntarily fails to act (omission). In this respect, the term 'recklessness' means indifference to the possible risks and consequences of one's conduct.

In other words, 'gross negligence' entails extreme and serious carelessness that shows willful or reckless disregard and represents a conscious violation of others' rights to safety[12] or to property.

Instead, 'willful misconduct' means any act or failure to act by a person that is intended to cause harm to others. In this case, a person is aware that a harmful outcome may occur as a consequence of his/her action and he/she wants to cause that consequence. It commonly means intentionally doing that which should not be done and intentionally wanting to cause damage to another.

In criminal law,[13] the definition of willful misconduct is different from common and civil law systems.

Common law systems have always drawn a sharp distinction between negligence, however gross, on the one hand, and 'wilful misconduct', on the other.[14]

In common-law systems, negligence is defined as a gross deviation from a reasonable standard of care. This is a higher standard than negligence under tort law. The level of care required is generally defined as that which would be exercised by a 'reasonable person'. The basis for accusing a person of criminal negligence is that he has some duty to exercise care towards a specific person or a whole class of people. He must then fail to meet that duty, and such a failure must cause personal injury.

Criminal-law negligence is variously defined, but according to a common legal definition, it is characterized by 'inadvertent taking of an unjustifiable risk'.[15]

The following leading cases, which are relevant in other areas characterized by risk-related activities, specify this definition and can help deal with the questions that arise in the field of aviation law.

In the 'R. v Adomako' case (1994),[16] the defendant, an anaesthetist, failed to notice that an oxygen tube became disconnected from the patient during a surgical

---

[11]Lyng (1990), p. 851 ff.

[12]Reason et al. (1994) and Hudson et al. (1997).

[13]Laudan (2006); Gooderham (2009), p. 358.

[14]The doctrine of the common law is that: "Gross negligence may be evidence of *mala fides*, but is not the same thing". So 'Goodman v. Harvey' (1836) 4 Accident & Emergency (A. & E.) pp. 870 ff.

[15]Ormerod (2005).

[16]'Regina v Adomako' (1994) 1 AC 171, Washington Law Review (WLR), 3, p. 238.

operation. As a result, the patient died from cardiac arrest. The doctor's incompetence was felt to have been so bad that it deserved to have criminal-law implications.

The defendant challenged the conviction on two grounds: on one hand, he argued that there was no breach of the duty of care, on the other hand, that there was not a sufficient causal link between his behaviour and the patient's death.

Following this case, a conviction for 'gross negligence' manslaughter requires the existence of a duty of care, and breach of that duty resulting in death or a risk of death which would be obvious to a reasonably prudent person. Additionally, the defendant's conduct must have fallen so far below the standard of a reasonable practitioner as to be grossly negligent in the view of the jury, and thereby warranting a criminal conviction for manslaughter.

The trial judge applied the four-stage test (so-called 'Adomako test'). The four elements of the offence are:

1. The defendant owed the victim a duty of care.
2. The defendant breached that duty.
3. The breach caused (or significantly contributed to cause) the victim's death.
4. The breach was grossly negligent.

The prosecution needs to demonstrate that there are all four elements.

In the *R. v Misra & Srivastava* case (2005)[17] the two appellant doctors were convicted of 'gross negligence' manslaughter following the death of a post-operative patient, who developed an undiagnosed infection in the wound, despite obvious symptoms. During the process, the defendants argued that the 'Adomako test' was not sufficiently clear and precise, as there is no definition of 'gross negligence'.[18] In particular, they therefore contested the incompatibility of this test with their rights under Article 7 of the European Convention of Human Rights (ECHR).[19]

This Article not only enshrines the principle *nullum crimen, nulla poena sine lege*, which prohibits criminal convictions without a legal basis, but also lays down the fundamental requirements of clearness and preciseness of criminal law, so as to enable individuals to ascertain which conduct constitutes a criminal offence and to be aware of the consequences of infringements of the law.

In the *Misra* case, the Court of Appeals first considered that manslaughter by 'gross negligence' did not conflict with the requirement of legal certainty under

---

[17]'R v Misra and Srivastava EWCA' (England and Wales Court of Appeal), Criminal Division, 2005, 328, p. 2375. For a commentary, see Erin and Ost (2007), p. 21; Allen (2007), p. 340.

[18]See Martin (2013), p. 79.

[19]Article 7(1) ECHR says: "No one shall be held guilty of any criminal offence on account of any act or omission which did not constitute a criminal offence under national or international law at the time when it was committed, Nor shall a heavier penalty be imposed then the one that was applicable at the time the criminal offence was committed". For a commentary see O'Neill (2001); Salinas de Frias (2013), chapter 5; Ormerod and Laird (2014), p. 11; Brems and Gerards (2014), p. 252; Bjorge (2015), p. 208; Martin and Storey (2015), p. 27 f.; Beauman and Wilson (2018), chapter 2, para. 2.3 (Fatal offences against the persons).

Article 7 ECHR. It also clarified, however, that only a risk of death (and not a risk of bodily injury) shall be considered as an offence in the health sector.

In particular, the Court held that the conduct of the defendant in performing professional obligations to the patient was "truly exceptionally bad" and showed a high degree of indifference to an obvious and serious risk to the patient's life.

In response to the criticism concerning the 'Adomako test', the Court stated:

> The decision whether the conduct was criminal is described not as 'the' test, but as 'a' test as to how far the conduct in question must depart from accepted standards to be 'characterised as criminal' [...] The question for the jury is not whether the defendant's negligence was, and whether, additionally, it was a crime, but whether his behaviour was grossly negligent and consequently criminal. This is not a question of law, but one of fact, for decision in the individual case.[20]

More recently, in the *R. v Zaman* case (2017),[21] the Court of Appeal stated that 'gross negligence' manslaughter is an area in which the sentencing exercise is particularly focused on the facts of the case. In this case, a man, who had a severe allergy to peanuts, ate a takeaway meal from an Indian restaurant. He was told by the waiter that the food was nut-free. Unfortunately, the sauce contained a significant amount of peanut, which led the customer to go into anaphylactic shock, which killed him. Therefore, on the basis of the facts, the Court confirmed a 6-year prison sentence handed down against the restaurant owner who had persistently failed to take measures to ensure that customers suffering from peanut allergies were not served with food containing peanuts. Thus the culpability was very high considering the very serious (fatal) consequences.

In this case, the Court also made clear that deliberately risking the death and safety of customers could result in a finding of 'gross negligence' manslaughter and a severe penal sentence.

The aforementioned leading cases demonstrate the difficulties that still exist in defining and differentiating the concepts of 'wilful misconduct' and 'gross negligence' from a legal point of view.

In conclusion, what does the term 'negligence' precisely mean in criminal law?

Gross (and, consequently, criminal) negligence is considered an act of omission or commission, where a person demonstrates the wilful disregard for the rights of other people that results in possible or actual harm.

It is a conscious, voluntary violation of rules or procedures that may cause serious personal injury to other people.

In the context of civil liability, instead, 'gross negligence' is a conscious and voluntary disregard of the need to use reasonable care, which is likely to cause foreseeable serious injury or harm to persons, or property.

Therefore, 'gross negligence' is more serious than 'mere negligence', which—as has been said—is simply a failure to exercise reasonable care. In fact, they differ not

---

[20]'R v Misra and Srivastava EWCA', cited above, para. 62.

[21]'R. v Mahammed Khalique Zaman', in EWCA (England and Wales Court of Appeal), Criminal Division, 2017, p. 1783. See Parker-Holland (2018).

only in the varying degree of carelessness or negligence, but also with reference to intent to cause damage. Therefore, the fault/failure analysis must be based on the 'reasonableness test' and on the assessment of foreseeable damage (risk avoidance).[22]

3.3 Towards a Common Legal Definition of 'Gross Negligence' in Aviation

In aviation industry, 'gross negligence' plays a significant role in technical decisions. In this sector, negligence may coincide with an operating error, which may be considered as inexcusable when strict rules and procedures have been violated.

An example of this is the Tuninter Flight 1153 disaster.[23] On August 6, 2005 an ATR-72, flying between Bari and Djerba (Tunisia), ditched into the Mediterranean Sea, about 18 miles from the city of Palermo, in Sicily, killing sixteen of the 39 people on board. The accident resulted from fuel exhaustion due to the installation of a fuel quantity indicator designed for the ATR-42 in the ATR-72. Both the captain and co-pilot (together to executives and technicians) were brought to court with charges for manslaughter, unpremeditated disaster and heavy injuries.

At first sight, the pilots should not have been charged, in accordance to 'just culture' principles, since the disaster occurred for a technical cause, beyond their control. They could not know, on the basis of their experience and training, that the fuel indicator was incompatible with that aircraft type and that it provided a wrong indication about the quantity of fuel in the fuel tanks.

Nevertheless, the investigation reports demonstrated with simulations, that the pilot would have been able to land at the Palermo airport if he had followed emergency procedures to decrease friction.

The Italian judges accepted the prosecutors' opinion that the pilots, instead of making a crash landing on the sea, should have been able to glide to Palermo airport. Thus the standard procedures had been breached by the pilots ('gross negligence'). They were sentenced to 10 years' imprisonment.

The concept of 'gross negligence' is acknowledged by both civil and criminal law.

Many international conventions on air carriers' civil liability,[24] inspired by the common law principles, put the two concepts of 'willful misconduct' and 'gross negligence' on a par.

---

[22]Hudson and Guchelaar (2003), pp. 98 ff.

[23]See Michaelides-Mateou and Mateou (2012), chapter 9 (Cases of Prosecution of Aviation Professionals); Dekker (2016), Chapter 4 (The Criminalization of Human Error), para. 1 and (2011), pp. 121 ff.; Suhir (2018), p. 155.

[24]Ivaldi (1990), p. 149; Zampone (1997), pp. 1287 ff.; Zampone (1999); Zampone (2001), p. 215.

For example, according to Article 22(5) of the Montreal Convention 1999[25] for the Unification of Certain Rules for International Carriage by Air, the carrier cannot limit their civil liability 'if it proved that the damage resulted from an act or omission' of himself, 'its servants or agents, done with intent to cause damage or recklessly and with knowledge that damage would probably result'.

In this context, an act or omission done with the intent to cause damage can be qualified as 'willful misconduct' (*dolus*), whereas an act or omission done recklessly and with knowledge that damage would probably result can be qualified as 'gross negligence' (*culpa lata*).

Although this international instrument conceptually distinguishes between 'willful misconduct' and 'gross negligence', it puts them on the same level with regards to the consequences.

The Warsaw Convention of 1929[26] was more explicit where it reads as follows:

> The carrier shall not be entitled to avail himself of the provisions of this Convention which exclude or limit his liability, if the damage is caused by his wilful misconduct or by such default on his part as, in accordance with the law of the Court seized of the case, is considered to be equivalent to wilful misconduct (Article 25(1)).

In this case, reference is made expressly to 'wilful misconduct' or to such default that is considered to be equivalent thereto, *i.e.* 'gross negligence'.

From a terminological point of view, it may primarily be interesting to draw attention to the definition contained in the Glossary of terms, attached to the document 'Explanatory material on ESARR 2 Requirement',[27] issued by Eurocontrol, where, consistently with the common meaning, 'gross negligence' is defined as "any action or an omission in reckless disregard of the consequences to the safety or property of another".

This technical-legal document, containing safety regulatory requirements, also defines '(ordinary) negligence' as follows: "where there is a duty of care and a person fails to exercise such care, skill or foresight as a reasonable person in that situation would exercise". The differences between the two concepts are based on the seriousness of unsafe conduct and are applicable to both civil and criminal proceedings.

It is interesting to recall the position of Eurocontrol on this subject. In its paper, entitled 'Just Culture Policy',[28] this Organisation notes that there does not seem to be a commonly agreed definition of 'gross negligence' in Europe.

---

[25] See Giemulla et al. (2017), pp. 2 ff.

[26] Convention for the Unification of Certain Rules Relating to International Carriage by Air, signed at Warsaw on 12 October 1929, as amended by the Protocol Signed at the Hague, on September 28, 1955. See Beumont (1955), pp. 414 ff.; Riguzzi (1974), pp. 248 ff.

[27] Eurocontrol, ESARR Advisory Material/Guidance Document EAM 2/GUI 4—Explanatory Material on ESARR 2 "Requirements on the implementation by States of an Occurrence Reporting and Assessment Scheme for Air Traffic Management (ATM)", 9th edition, 9 August 2004, edition 1, Annex 1, Glossary, 33.

[28] Document that contains the conclusions of the Provisional Council at its session of 10th May 2012. Eurocontrol (ed.), Just Culture Policy, that endorsed a 'Model Policy regarding criminal

Nevertheless "It seems to be however generally agreed that gross negligence implies a degree of severity, serious disregard to an obvious risk and profound failure to take such care that is evidently required in the circumstances".[29]

This statement, which contains another definition of 'gross negligence', similar to its previous one, is significant as it represents the first valuable effort to provide more clarity on this subject.

More recently, unsuccessful attempts to introduce a definition of 'gross negligence' in European legislation have been made.

It should be recalled that the proposal[30] for Regulation No. 376/2014 of the original text, as drawn up by the Commission, also contained a definition of 'gross negligence', a concept closely linked to that of 'just culture'.

Taking into consideration that this proposal contained—as we have seen—provisions to promote the appropriate environment for encouraging aviation professionals to report safety related information by protecting them from punishment, except in cases of 'gross negligence', a clear definition of this concept was necessary to overcome the current interpretative uncertainties.

Article 2(4) of this proposal stated as follows: "'gross negligence' means a manifest and willful violation of the duty of care directly causing foreseeable damage to a person or to a property, or which seriously lowers the level of aviation safety".

Nevertheless, this definition was deleted from Regulation No. 376/2014 because an agreement among the Member States on the wording of this proposed provision had not been reached. This was due to differences between common law and civil law systems, in spite of the firm intention, expressed by the Commission, to maintain such as definition in the final text. Therefore, the matter has now been temporarily shelved.

The Commission's view is completely justified[31] taking into account that, in order to apply 'just culture' principles, a clear division line between acceptable ('mere negligence') and unacceptable ('gross negligence') behaviour must be drawn.

The above mentioned definition took a slightly different view compared to the previous, elaborated by Eurocontrol, because it did not make any reference to the element of the reckless disregard of the consequences and, instead, was referred to a willful violation of the duty of care.

---

investigation and prosecution of criminal offences further to the reporting of civil aviation incidents and accidents', elaborated by the Just Culture Task Force (JCTF), Brussels, September 2012, p. 5.

[29]See the previous footnote.

[30]See COM (2012)776 final, Brussels, 18 December 2012: "Proposal for a Regulation of the European Parliament and of the Council on occurrence reporting in civil aviation amending Regulation (EU) No. 996/2010 and repealing Directive No. 2003/42/EC, Commission Regulation (EC) No. 1321/2007 and Commission Regulation (EC) No. 1330/2007", Article 2(4).

[31]*Contra* van Dam Roderick during the Seminar "Just Culture in the context of occurrence reporting schemes" Brussels, 19 April 2012.

Considering that the term 'willful', used in the present definition, is mainly understood as 'intentional' and suggests 'intentional misconduct', it would have been preferable to refer to the word 'reckless'.

However, in my view, the intention inherent in the term 'wilful' is, in this definition, referred to the violation of the duty of care and not also to the criminal intent to cause damage.

In fact, it should be noted that 'willful misconduct' is a conduct that presupposes a deliberate intention to cause damage, whereas 'gross negligence' is not characterized by intention to cause damage, but by serious carelessness (wilful and manifest violation of the duty of care), which causes foreseeable damages or significantly reduces the level of 'aviation safety'.

In this context, the aforementioned definition of 'aviation safety', contained in ICAO Annex 19 ('the state in which risk associated with aviation activities, related to, or in direct support of the operation of aircraft, are reduced and controlled to an acceptable level') should be recalled.[32]

This equivalence between the two definitions of 'gross negligence' (deleted from the text of Regulation No. 376/2014) and 'safety' (contained in ICAO Annex 19) highlights the following principle: any action that significantly and seriously affects aviation safety, lowering its level, must be considered unacceptable.

In my opinion, this definition should not be scrapped from the final text of Regulation No. 376/2014, but, at best, its wording needs be modified in order to delete the reference to the unclear expression 'wilful violation' and replace it with the following wording 'reckless disregard of the consequences' or something similar.

It is important, however, to point out that, even though the definition of 'gross negligence' has been removed from Article 2 of Regulation No. 376/2014, it has been implicitly maintained in its recital 37 and also in Article 16(10).

In fact, recital 37 highlights that aviation operators should not be subject to any prejudice on the basis of information reported, except in cases of 'willful misconduct' or 'where there has been manifest, severe and serious disregard with respect to an obvious risk and profound failure of professional responsibility to take such care as is evidently required in the circumstances'.[33] It refers to serious disregard for a foreseeable risk and severe failure to take professional care (due diligence) as circumstances required.

Therefore, we can consider it another definition of 'gross negligence'.

Since the last examined notion is contained in a recital of a EU Regulation, and not in Article 2 (entirely devoted to definitions), attention can be drawn to the legal nature of recitals of a EU Act, located in the preamble, between the title and its

---

[32]Chapter I of the ICAO Annex 19, cited above.

[33]In the legislative proposal COM (2012)776, in fact, the corresponding recital 39, in this connection, referred to the case of 'gross negligence', without feeling the need to define it, given that its definition was included in the following Article 2(4).

enacting terms.[34] The preamble is composed by a numbered list of recitals, containing objectives, principles, references to other legislations and occasionally definitions. Recitals do not contain binding previsions and do not have the normative status of the enacting terms, which constitute the normative part of the act and are divided into articles. Even though recitals are non-binding, they are an essential component in the legal interpretation of the act.

Therefore, regarding the specific case under examination, when articles of Regulations No. 376 refer to 'gross negligence', in the absence of a definition of this concept in the normative part, it should be interpreted as having the meaning indicated in the preamble.

Actually, in this case the aforementioned condition of the lack of a clear definition of 'gross negligence' in the normative part of Regulation 376 is not fulfilled.

In fact, Article 16(10), defines a concept that differs from 'wilful misconduct', namely that of a 'manifest, severe and serious disregard of an obvious risk and profound failure of professional responsibility to take such care as is evidently required in the circumstances, causing foreseeable damage to a person or property, or which seriously compromises the level of aviation safety'. The latter definition, in the first part, recalls almost to the letter what is contained in recital 37, but in the second part ("causing foreseeable damage to a person or property, or which seriously compromises the level of aviation safety"), concerning the consequences, replicates the wording already contained in the definition of the original proposal.[35]

These two definitions, respectively contained in recital 37 and in Article 16(10), are more like that issued by Eurocontrol, rather than with that already contained in the original legislative proposal.

Despite the different wording between the two current provisions in respect to the removed definition, contained in the original proposal, we may note that there is a lot of common ground between them. In fact, the expression 'profound failure of professional responsibility to take such care' can be considered substantially equivalent to the expression "a manifest and willful violation of the duty of care". In addition, the concept of 'manifest, severe and serious disregard with respect to an obvious risk" is not much different from an action "directly causing foreseeable damage".

In all these definitions appears the concept of conscious and voluntary disregard of the need to use reasonable care, able to cause foreseeable and serious injury.

However, if 'gross negligence' means serious disregard of, or indifference to, an obvious risk, 'ordinary negligence' is a mere failure to exercise reasonable care. An act of 'mere negligence' is an unintentional breach of duty of care and leads to unintended harmful consequences.

The difficulty in defining and understanding the meaning of legal expressions such as 'gross negligence' and '(ordinary) negligence' causes much confusion, as

---

[34]So Humphreys et al. (2015), pp. 41 ff. See also Sparck-Jones (1972), pp. 11 ff.; Klimas and Vaiciukaite (2008), p. 2 ff.

[35]Article 2 (4) COM (2012)776.

there is no agreement on an uniform definition from country to country at the European and international levels.

In any case, a new definition of 'gross negligence' should be created, in line with a non-punitive (or 'just culture') approach and compatible with the definition drawn up by Eurocontrol, mentioned above.

In order to eliminate all uncertainties, a clear distinction between 'mere negligence' and 'gross negligence' in the field of aviation should be defined by the EU legislature.

In fact, among the measures that must be taken in order to draw a sharp dividing line between acceptable and unacceptable behaviour, a pivotal role is certainly played by a clear legal definition of 'gross negligence'. This definition avoids questionable legal decisions, which attribute the responsibilities of front-line operators to be greater than their professional duties.

References

Allen MJ (2007) Textbook on criminal law, 9th edn. Oxford University Press, Oxford

Baumgartner M (2011) The role for the human being in single European Sky. In: Calleja CD, Mendes de Leon P (eds) Achieving the Single European Sky: goals and challenges. Kluwer, Alphen aan den Rijn, pp 299–314

Baumgartner M, Chalck M, Pidgen J (2011) Just culture. In: Calleja CD, Mendes de Leon P (eds) Achieving the Single European Sky: goals and challenges. Kluwer, Alphen aan den Rijn, pp 281–298

Beauman C, Wilson C (2018) My revision notes: OCR a level law. Hodder Education, Hachette UK Company, London

Beumont KM (1955) Warsaw Convention of 1929, as amended by the protocol signed at the Hague. J Air Law Commerce 22(4):414–433

Bjorge E (2015) Domestic application of the ECHR: courts as faithful trustees. Oxford University Press, Oxford

Brems E, Gerards J (eds) (2014) Shaping rightd in the ECHR: the role of the European Court of human rights in determining the scope of human rights. Cambridge University Press, Cambridge

Dekker S (2011) The criminalization of human error in aviation and healthcare: a review. Safety Sci 49:121–127

Dekker S (2016) Just culture: restoring trust and in your accountability organization, 2nd edn. CRC Press, Taylor & Francis, Abingdon-on-Thames

Duguid B (2006) What makes negligence 'Gross' and when is misconduct 'Wilful'? blakes bulletin on energy - oil & gas. Cassels & Graydon LLP (Blakes), Toronto

Duval C, LeLeuch H, Pertuzio A, Weawer J (2009) International petroleum exploration and exploitation agreements: legal, economic & policy aspects, 2nd edn. Barrows Company, New York

Erin CA, Ost S (eds) (2007) The criminal justice system and health care. Oxford University Press, Oxford

Giemulla E, Schmid R, Müller-Rostin W, Dettling-Ott R, Margo R (2017) Convention for the unification of certain rules for international carriage by air. In: Giemulla E, Schmid R (eds) Montreal convention. Kluwer Law International, BV, Alphen aan den Rijn, pp 1–22

Gooderham P (2009) The distinction between gross negligence and recklessness in English criminal law. J Royal Soc Med 102(9):358

Grosse A (2010) Theroux M.P., Gross Negligence: how bad does it have to be? Bennett Jones LLP, Calgary

Hudson PTW, Guchelaar H-J (2003) Risk assessment in clinical pharmacy. Pharmact World Sci 25 (3):98–103

Hudson PTW, Verschuur WLG, Lawton R, Parker D, Reason J (1997) Bending the rule II: the violation manual. Rijks Universiteit, Leiden

Humphreys L, Santos C, di Caro L, Boella G, van der Torre L, Robaldo L (2015) Mapping recitals to normative provisions. Jurix 2015: The Twenty-Eighth Annual Conference, Braga, 10-11 December 2015. In: Rotolo A (ed) Legal knowledge and information systems: proceedings. IOS Press, Amsterdam, pp 41–50. 278

Ivaldi P (1990) Diritto uniforme dei trasporti e diritto internazionale privato. Giuffrè, Milano

Klimas T, Vaiciukaite J (2008) The law of recitals in European community legislation. ILSA J Int Comp Law 15:1–33

Laudan L (2006) Truth, error and criminal law: an essay in legal epistemology. Cambridge University Press, Cambridge

Lyng S (1990) Edgework: a social psychological analysis of voluntary risk taking. Am J Social 95 (4):851–856

Martin J (2013) Key facts criminal law, 4th edn. Routledge, Abingdom, p 79 ff

Martin J, Storey T (2015) Unlocking criminal law, 5th edn. Routledge, Abingdon

Michaelides-Mateou S, Mateou A (2012) Flying in the face of criminalization: the safety implications of prosecuting aviation professionals for accidents. Routledge, Abingdon-on-Thames

O'Neill A (2001) EU law for lawyer. Hart Publishing, Oxford

Ormerod D (2005) Smith & Hogan's criminal law, 11th edn. Oxford University Press, Oxford

Ormerod D, Laird K (2014) Text, cases, and materials on criminal law. Oxford University Press, Oxford

Parker-Holland J (2018) A recipe for disaster? Gross negligence manslaughter in R vs. Zaman. In: Keep Calm and Talk Law, www.keepcalmtalklaw.co.uk

Reason J, Parker D, Free R (1994) Bending the rules: the varieties, origins and management of safety violations. Rijks Universiteit, Leiden

Riguzzi M (1974) In tema di colpa temeraria e consapevole prevista dall'art. XIII del Protocollo dell'Aja del 1955. Diritto aereo:246–264 ff

Rodhes L (2013) Limitations on liability exceptions for gross negligence and willful misconduct and the implications for outsourcing agreements. Bus Technol Sourcing Rev:19

Salinas de Frias AM (2013) Counter-terrorism and human rights in the case of the European Court of Human Rights. Council of Europe, Strasbourg

Sparck-Jones K (1972) A statistical interpretation of term specificity and its application in retrieval. J Doc 28(1):11–21

Suhir E (2018) Human-in-the-loop. Probabilistic modeling of an Aerospace Mission Outcome. CRC Press/Taylor & Francis Group, Boca Raton

Sweeney DH (2016) Oil and gas joint operating agreements: a Comparative world-wide analysis. LexisNexis, New York

Zampone A (1997) La condotta temeraria e consapevole nel diritto uniforme dei trasporti: gli elementi caratterizzanti. In: Studi in onore di G. Romanelli. Giuffré, Milano, pp 1287–1324

Zampone A (1999) La condotta temeraria e consapevole nel diritto uniforme dei trasporti. Giuffrè, Milano

Zampone A (2001) Sulla nozione di wilful nisconduct nella giurisprudenza statunitense alla luce dell'entrata in vigore del Protocollo di Montreal n. 4. Diritto dei trasporti 1:215–219

Chapter 4
'Just Culture' *Versus* 'Blame Culture' in Aviation

4.1 The Differences Between 'Blame Culture' and 'Just Culture' in Aviation

As we have seen, the opposite of a blame culture is a culture where people feel free to admit their errors and make suggestion.

In the aviation sector, the so-called 'non-punitive culture' is in conflict with the traditional 'blame culture' (also called 'punitive culture').

Some legal systems require 'gross negligence' to justify criminalisation, others permit punishment also for 'simple negligence'.

The latter solution adopt a 'blame culture' approach, aimed at criminalizing all types of negligence[1] in order to ensure deterrence, in other words to discourage people from behaving illegally.

The punitive culture has been perceived by operators to be unjust and ineffective in recent years.

In fact, a 'culture of punishment', which pursues all types of infringements, makes no distinction between conscious error, which is the deliberate breaking of a rule (violation) and unintentional error (a genuine mistake), treating both in the same way. Although many such errors are unintentional, the front-line operator can nevertheless be liable for acts of 'mere negligence'.

In any case, changing to a law that is inspired by non-punitive cultural principles and avoids any indiscriminate punishment would be welcomed.

According to a 'just culture' approach, if a person was inadvertently negligent in doing something or honestly did not comply with generic or specific rules, appropriate to the matter involved, he should not be punished. On the contrary, if a person can envisage and predict the negative consequences of his/her behaviour and actions

---

[1]Coelho dos Santos (2013), pp. 40 ff.

© Springer Nature Switzerland AG 2019
F. Pellegrino, *The Just Culture Principles in Aviation Law*, Legal Studies in
International, European and Comparative Criminal Law 3,
https://doi.org/10.1007/978-3-030-23178-1_4

and it is sure that he/she was able to avoid the harmful outcome, he/she should be punished. It is the element of consciousness that makes behaviour unacceptable.

The contrast between 'just culture' and 'blame culture' reflects the different approach to aviation safety, pursued through safety/technical investigation, on the one hand, and judicial investigation for aircraft accidents and incidents, on the other. The former one is aimed at preventing the risk of future accidents by taking appropriate measures to avoid a repetition of such as events. The latter is aimed to prosecute those who are responsible.

While the aim of technical investigations is to determine the causes of an accident or serious incident in order to avoid similar events in the future and to increase safety standards by following a proactive approach,[2] judicial inquiries are aimed at prosecuting people found guilty or those believed to be responsible, by adopting a repressive approach. The former is conducted with the sole intention of identifying the causes of the accident and making safety recommendation to prevent the recurrence of similar events, the latter aims to apportion blame or liability.

The sole aim of the technical investigation is to prevent future risks, whereas the task of judges consists primarily of ascertaining whether the wrongful conduct of frontline operators is a *conditio sine qua non* (logical antecedent) for a damaging event. To this end, the judicial inquiry has to verify if the conduct is due to a breach of laws or any regulation, order or instruction or is characterized by 'negligence', 'imprudence' or 'lack of skill'.[3]

Another issue that should be stressed is the lack of legal certainty. Aviation professionals may find themselves faced with criminal offences in one country, and not in another.[4] In addition, they might not be sure that statements made by them during the technical investigation could be used in any subsequent judicial proceedings.

In the face of errors that lead to accidents, the traditional solution is to blame the operators involved. This criminalization[5] approach is unfair and inadequate, considering that front-line personnel is required to take delicate decisions. For instance, air traffic controllers,[6] as guarantors of aviation safety, to assist flight crew and to intervene in critical situations, are required to comply with strict duties that often go beyond provisions contained in the technical rules and regulations. That situation very easily exposes them to both civil and criminal prosecution.

---

[2]Dusi (2017), pp. 35 ff.

[3]See Article 43 of the Italian Criminal Code.

[4]See Directive 2014/41/EU of 3 April 2014 regarding the European Investigation Order in criminal matters (O.J. L 130 of 1 May 2014).

[5]Gates (2009), pp. 9 ff.; Trögeler (2010), p. 36; Trögeler (2011), pp. 2 ff.; Michaelides-Mateou and Mateou (2012), chapter 7 (The Effect of Criminalization on Aviation Safety); Nemsick and Passeri (2012), p. 21.

[6]See Franchi (2011), pp. 73 ff.

This approach is insufficient, considering the very narrow time margins that are required for pilots and ATCs to take delicate decisions.[7]

Therefore, by following the traditional, punitive approach, even if operators comply with numerous legal/technical provisions and very complex procedures, and even if they conduct their job with undisputed professionalism, there is no guarantee they would be able to avoid criminal prosecution.[8]

In addition, main studies found that punishing air traffic controller or pilots has led to a reduction in the reporting of incidents and sharing of safety information.[9] It should also be noted that the need for a culture that encourages honest errors reporting is not yet "reconciled with the judicial system and legislators".[10]

There can be no doubt that guilt and blame are considered as obstacles to transparency,[11] which is necessary to prevent further adverse events. In fact, they inhibit the reporting of an aviation incident, making it difficult to identify the appropriate corrective measures to handle new risks to aviation safety.

Aviation is characterized by such a high level of professionalism and complexity that certain corrective measures are required.

Within the 'just culture' approach, error is considered as a possible event: therefore an assessment of its acceptability is needed.

Risk prevention is pursued not purely by the fulfillment of legal obligations, regulations or technical standards, but also by personal considerations based on training and on-the-job experience. There are non-rational factors which are based on personal perception, culture, emotion, psychological attitude, individual experience, lifelong habits etc. The so-called 'human factor'[12] includes all of these elements and plays a special role in the safety culture framework.

Nevertheless, as has been observed, it should not be forgotten that the judges conduct an *ex post* evaluation according to principles of law and mostly ignore the highly technical competences and performances of front-line operators.[13]

ICAO defines competency as the combination of the knowledge, skills and attitudes required to perform a task to a prescribed standard.[14] Therefore, competency is taken to mean possession of the required level of knowledge, skill,

---

[7]Orasanu (1993), pp. 137 ff.; Weiner et al. (1993); Murray (1997), pp. 83 ff.

[8]See Barra (2013), pp. 8 ff.

[9]See Fassert (2005); Dekker and Laursen (2007), pp. 50 ff.; Kodate and Dodds (2008), p. 10.

[10]See SAFREP (Safety Data Reporting and Data Flow Task Force), Report on ATM Incident Reporting Culture: Impediments and Practices, Issue 1.0, ref. DAP/SAF/126 of 13 October 2005, p. 5.

[11]Fassert (2001).

[12]See Del Pinto (2017).

[13]See Dekker (2013a), p. 8. At this purpose he says that judges: "do not know the messy details, they lack technical knowledge and misunderstand the subtleties of what it takes to get the job done despite the organization, the rules, the multiple constraints". See also Marx (2012).

[14]So Procedures for Air Navigation Services—Training (PANS-TRG, Doc 9868, published on 10 November 2006), Chapter 1 (definitions and acronyms).

experience[15] (and, when required, proficiency in English), to permit the safe and efficient provision of aviation services.[16]

Other studies have clearly demonstrated that human error,[17] in most cases, is not just an individual error, but also an organizational one.[18] The problem is rarely the fault of an individual, often it is the fault of the system as a whole.

For example, in both the Milan Linate[19] and Überlingen[20] disasters, instead of a failure of front-line operators, shortcomings in the organizational system[21] due to the management[22] were the leading cause of the disaster. Both accidents led to fierce retribution against individuals even though hidden systemic problems were the cause.

4.1.1 The Linate Disaster

The Linate disaster, the most serious accident in Italian aviation history, occurred on 8 October 2001 at Milan Linate airport, when a McDonnell Douglas MD-87 aircraft,

---

[15]DeWinter (2013), p. 38.

[16]See Blajev (2013), pp. 6 f.

[17]Salaman (1992); Garland et al. (1999); Shorrock (2013), pp. 32 ff.

[18]Weick (1987), pp. 112 ff.; Cooper and Robertson (1988); Rousseau (1988), pp. 139 ff.; Schein (2010), pp. 35 ff.; Bate (1992), p. 229; Weick and Roberts (1993), pp. 357 ff.; Hofstede (1991); Reason (1995), pp. 1708 ff.; Thompson et al. (1996), p. 647 ff.; Westrum and Adamski (1999), p. 67 ff.; McDonald et al. (2000), pp. 51 ff.; Dekker (2013b), p. 38.

[19]The Linate disaster occurred on 8 October 2001 at Linate Airport in Milan, when Scandinavian Airlines Flight 686, a McDonnel Douglas MS-87 airliner, carrying 110 people, bound for Copenhagen, collided on take-off with a Cessna Citation CJ2 business jet, carrying four people, bound for Paris. All 114 people on both aircrafts were killed, as well as four people on the ground. The Cessna was instructed to taxi from the western apron along the northern taxiway (taxiway R5) and then via the northern apron to the main taxiway which runs parallel to the main runway. Instead, the pilot taxied along the southern taxi route (taxiway R6), crossing See Cattaneo et al. (2006), p. 440.

[20]On the night of 1 July 2002, Bashkirian Airlines Flight 2937, a Tupolev Tu-154 passenger jet, and DHL Flight 611, a Boeing 757 cargo jet collided in mid-air over Überlingen, a southern German town on Lake Constance. All 69 passengers and crew aboard the Tupolev and the two crew members of the Boeing were killed. The two aircraft were flying at flight level 36,000 feet on a collision course, The airspace was controlled from Zürich by the private Swiss airspace control company Skyguide, The only air traffic controller handling the airspace was working two work-stations at the same time. He did not realise the problem in time and thus failed to keep the aircraft at a safe distance from each other, Less than a minute before the accident he realised the danger and contacted Flight 2937, instructing the pilot to descend by a thousand feet to avoid collision. After the Russian crew initiated to descend, their traffic collision avoidance system (TCAS) instructed then to climb, while at about the same time the TCAS on Flight 611 instructed pilots to of that aircraft to descend, Both aircraft followed those automated instructions and the collision occurred. All 69 passengers (including 57 children) and crew aboard the Tupolev and the two crew members of the Boeing were killed. See Bennet (2004), pp. 31 ff.

[21]Martorell et al. (2014), pp. 63 ff.

[22]Hayward and Lowe (2000); Panelli and Scarabello (2013), pp. 59 f.

carrying 110 people from Milan to Copenhagen, operated by Scandinavian Airlines, collided on take-off with a private Cessna business jet, carrying four people. All 114 people on board were killed, as well as another four people on the ground in the hangar.[23]

The findings of the experts appointed during the judicial investigation were that the main cause of the accident was the "runway incursion in the active runway by the Cessna" (human error of the pilot). However, the report held that this error was induced 'by a lack of experience and training of the pilot, by the use of an ambiguous wording, by a misleading airport marking system and by an inadequate cartography of the airport'.

Therefore, "human factor-related actions of the Cessna crew – during low visibility conditions – must be weighed against the scenario that led to the chain of events that caused the fatal collision; equally it can be stated that the system in place at Milano Linate Airport was ambiguous and inadequate regarding operational procedures".

In reference to the aerodrome operating conditions, the inquiry report also stressed that "the Milan Linate airport did not meet the safety standards specified in ICAO Annex 14; this is of particular importance given the fact that Linate is considered an important international airport".

Moreover, the report highlighted that, "while the Cessna crew made the original error, basic inadequacies in airport signage and poor communication meant that opportunities to correct it were missed".[24]

Investigation revealed the fatal collision was caused by significant malfunctions, technical shortcomings and sub-standard systems and procedures. Notably, at the time of the accident, the ground radar system was not working.

The same report also recognised the error of the controller, who, being unfamiliar with the existing airport sign age, did not notice the wrong location of the Cessna.

However, at the time of the accident, the controller was handling a heavy workload. In addition, due to the poor visibility (less than 200 m) and without of a radar system to determine the exact position of the aircraft, he was forced to trust the position reports provided by the Cessna's pilots. The controller detected no anomalies or potential risk in the position reported by the pilots. Therefore he instructed them to continue taxiing as he believed on good faith that the aircraft was moving along the right taxiway (R5).

During the first-degree trial, the victims' relatives exhibited the ANSV report[25] on the causes of aviation accidents, that highlighted relevant deficiencies in the airport's organizational system, but the defense lawyers protested against the exhibition of this document. The Tribunal and the Appeal Court became convinced that the conclusions of this report on the possible causes of the disaster could not be used as incriminating evidence against the accused. The defense addressed the same

---

[23]For technical aspects, see Garzone and Archibald (2010), pp. 213 f.

[24]So Kaminski-Morrow and Learmount (2004).

[25]ANSV, Relazione d'inchiesta, Incidente occorsi agli aeromobili Boeing MD-87, marche SE-DMA e CESSNA 525-A, marche D-IEVX, Aeroporto Milano Linate 8 ottobre 2001.

objections to the Supreme Court, which, however, did not accept them. Therefore, the report was finally admitted to the case file, under Article 234 of the Italian Criminal Procedure Code.

But nevertheless, criminal charges were levied against the airport traffic controller,[26] neglecting the other parties responsible for organizational problems.[27]

In fact, the final decision of the Supreme Court of Cassation of 5 June 2008,[28] condemned, *inter alia*,[29] the controller for multiple manslaughter (Article 589 of the Italian Penal Code) as well as negligent disaster (Articles 428 and 449 of the same Code).[30] He was sentenced to 3 years in prison for having not understood (actually, guessed) the Cessna's pilots confusion and disorientation, that led them to go along the wrong taxiway.

In light of this, it is clear that 'just culture' principles were not applied by the Court in respect to the controller's position.

However, this ruling, adhering to the view of the serious defects of the "airport system",[31] roughly confirmed the judgment of the Court of Appeal of Milan of 7 July 2006,[32] but failed to recognise a "position of guarantee" to the Airport Director. In fact, the Supreme Court noted that after Italian Legislative Decree No. 250 of 25 July 1997[33] entered into force to establish the ENAC, the Legislature had not conferred

---

[26]See Simpson (2014), part 2.

[27]There is not too dissimilar the Los Angeles runway disaster. On the evening of February 1, 1991, USAir Flight 1493, a Boeing 737-300, accidentally collided with SkyWest Flight 5569, a Metroliner turboprop aircraft, upon landing at Los Angeles. As Flight 1493 was on final approach, the local controller was distracted by a series of abnormalities, including another aircraft that inadvertently switched off the tower frequency. Thus the SkyWest flight was told to taxi in the take-off position while the USAir flight was landing on the same runway. The 737 collided with the twin-engine turboprop. All 12 people on board the smaller plane were killed, as well as 23 of the 89 passengers on the Boeing. The pilot of the Flight 1493 reported that he did not see SkyWest 5569 until he lowered the nose of his aircraft into the runway after landing The NTSB (National Transportation Safety Board) found that the probable cause of the accident was the procedures in use at the control tower, which provided inadequate redundancy, leading to a loss of situational awareness by the local controller. Therefore, NTSB recommended the use of different runways for take-off and landing. Nevertheless, the aircraft controller who put a jet and a commuter plane on the same runway, causing a fatal crash, was relieved of duty and condemned.

[28]Italian Supreme Court of Cassation, criminal division, IV section, No. 22614 of 5 June 2008, in www.comitato8ottobre.com.

[29]The Court of Cassation upheld a six-and-a-half year prison sentence for the former head of air traffic control authority ENAV. The former ENAV director general sentence of 4 years and 4-months sentence was confirmed. Former officials with the SEA airport agency had their sentences of 3 years and 3 months each upheld. The Tribunal let stand the Milan Court's acquittal of the former overall head of the Linate and Malpensa airports.

[30]For a commentary see Deiana (2006), pp. 52 ff.

[31]Camarda (2002); Sirena (2014), p. 6.

[32]Italian Court of Appeal of Milan of 7 July 2006, in www.comitato8ottobre.com On 16 April 2004, the Milan Tribunal found four persons guilty for the disaster: airport director and air-traffic controller, both sentenced to 8 years in prison, and former head of the air traffic control agency (ENAV) who received sentences of six and a half years.

[33]In G.U. No. 177 of 31 July 1997. For a commentary, see Di Palma (2000), pp. 11 ff.; Masutti (2010), pp. 473 ff.; Magnosi (2010), pp. 313 ff.

specific powers concerning regulation and supervision on the safe movements of aircraft inside the airport[34] to the airport managers.

4.1.2 The Überlingen Disaster

The Überlingen disaster raises similar issues. This tragedy was also due to systemic causes in the interplay between people, technology and procedures.

On 1 July 2002, a Tupolev 154 passenger jet, operated by Bashkirian Airlines, en route from Moscow to Barcelona, and a Boeing 757 cargo jet, operated by DHL and flying from Bergamo to Brussels, collided in mid-air (34,890 feet) over Überlingen, a southern German town on Lake Constance.[35]

At the time of the accident, both aircrafts were under the control and supervision of the Zürich ACC (Area Control Center), in accordance with an agreement between the German and Swiss ANSPs (Air Navigation Service Providers).[36] The radar system was operating only on fall-back mode and the main telephone system of connection with other ACC was shut down. During that night, according to the routine at Zürich ACC, only one of the two controllers on duty had worked. Systemic failings of management systems led to the air traffic controller failing to act in a timely manner. Therefore, traffic alert and collision avoidance systems were activated automatically. In fact, both aircraft had the Traffic Alert and Collision Avoidance System (TCAS)[37] installed. Indeed, this accident is the world's first

---

[34]De Meo (2000), pp. 763 ff.

[35]For a commentary see Schmitt et al. (2006); Hempel Lindøe et al. (2014), p. 72; Masys (2016a, b), pp. 71 ff.

[36]Control of the German airspace in which the accident occurred was delegated to the Swiss Air Navigation Services. In accordance with the requirements of ICAO Annex 11 (Chapter 3.5) the conditions of the delegation were established between the responsible Air Navigation Services in "Letters of Agreement" (LoA).

[37]TCAS (Traffic Alert and Collision Avoidance System) is a specific implementation of the ACAS (Airborne Collision Avoidance System) concept. ICAO Standards provide the general regulatory context for ACAS, while ICAO Procedures provide specific parameters in the operation of the systems. These two sets of documents establish that the pilot-in-command has full authority and full responsibility to select the course of action that will best resolve a traffic conflict and avert a collision, including the use of ACAS. ICAO Annex 2 (Rules of the Air), tenth edition, July 2005, section 3.2 specifies "Nothing in these rules shall relieve the pilot-in-command of an aircraft from the responsibility of taking such action, including collision avoidance maneuvers based on resolution advisories provided by ACAS equipment, as will best avert collision" and section 3.2.2, states "The aircraft that has the right-of-way shall maintain its heading and speed". The only implementations that meet the ACAS II standards set by ICAO are Versions 7.0 and 7.1 of TCAS II. ACAS II provides "Resolution Advisories" (RA's) in the vertical sense (direction) telling the pilot how to regulate or adjust his vertical speed so as to avoid a collision. According to Item 12 of page E-10 of ICAO's Proposed ACAS Performance-based Training Objectives, "if pilots simultaneously receive instructions to maneuvers from ATC and an RA which are in conflict, the pilot should follow the RA". For technical aspects, see Croucher (2015), pp. 2 ff.

midair collision with both aircraft having operating airborne collision avoidance systems.

Fifty seconds before the collision, the TCAS warned the Tupolev's crew about a collision course and forty seconds before the impact the controller became aware of the risk and instructed the Tupolev's crew to descend. The Russian aircraft began its landing, but at the same time its TCAS generated a contrary Resolution Advisory (RA) ordering it to climb. In spite of the TCAS order, the Tupolev continued its landing. Then this passenger aircraft collided with the DHL aircraft, which instead did not follow the air traffic controller instructions, but complied with its opposite TCAS Resolution Advisory. All 71 occupants (51 children) of the two aircraft were killed in the collision.

A technical investigation was carried out by the German Federal Bureau of Aircraft Accident Investigation (BFU).[38] Its official report[39] identified a number of failings in both the air traffic control centre and in the training and use of collision avoidance systems.[40]

As the simulator showed, the TCAS had generated instructions that would have led to a sufficient vertical separation of both aircraft. In other words, if both aircraft had followed these automated instructions, the collision would not have occurred. In fact, no proper lessons had been taken from a near-miss which occurred about a year before this accident in Japanese skies.[41] In that case, disaster was avoided because both pilots followed the TCAS instructions, ignoring the controller's orders.

Immediately following the accident it was argued that pilots should have known the importance of following the TCAS, irrespective of what the controller stated.[42]

Rules on the TCAS are contained in many ICAO documents, published prior to 1 July 2002 (Annex 2,[43] Annex 10[44]; Annex 11[45]; Doc 8186[46]; Doc 4444[47]). In

---

[38]BFU is acronym for *Bundesstelle für Flugunfalluntersuchung*.

[39]BFU, Investigation Report, Braunschweig, AX001-1-2/02 May 2004, 1 July 2002, pp. 34 f.

[40]See Turney (2007), pp. 2 ff.

[41]On 31 January 2001, two Japanese aircraft (Boeing 747 and McDonnell Douglas DC-10), both operated by Japan Airlines, nearly collided which each other in Japanese skies. Also the two planes were on a collision course and both planes had received conflicting instructions from their TCAS and the controller.

[42]So Johnson (2008).

[43]Chapter 3.

[44]Aeronautical Telecommunication, Vol. IV, Surveillance Radar and Collision Avoidance Systems, Attachment A.

[45]ICAO Annex 11 to the Convention on International Civil Aviation, Air Traffic Services, Thirteenth Edition, July 2001. It is now into its Fourteenth Edition, July 2016.

[46]PANS-OPS, Procedure for Air Navigation Services—Aircraft Operations, Vol. I, Flight Procedures, Doc 8168, OPSfirst ed. 2006, Attachment to Part III, Section 3, Chapter 3 ACAS Training Guidelines for Pilots.

[47]ICAO Doc 4444-ATM/501, Air Traffic Management. Procedures for Air Navigation Services, Fourteenth Edition, 2001, so-called ICAO PANS ATM (PANS is an acronym for Procedures for Navigation Services and ATM is an acronym for Air Traffic Management). It is now into its Sixteenth Edition, 2016 and defines the procedures for the provision of air navigation services.

particular, according to ICAO Doc 8168, if pilots simultaneously receive instructions to maneuvers from air traffic control (ATC) and TCAS Resolution Advisory (RA), which are in conflict, the pilot should follow the RA.[48] This means that, in case of conflict between TCAS RA and air traffic controller, the TCAS instructions always takes precedence. The reason for this statement is that ATC may not be aware of the TCAS RAs and may therefore issue opposite instructions.

On 4 September 2007, the Swiss District Court of Bülach (Switzerland)[49] found three managers of the air navigation service provider (Skyguide) to be guilty of (gross) negligent manslaughter, because they had failed to remedy well-known safety deficiencies and adequately supervise a safety-critical system.[50] They were sentenced to 12 months imprisonment and elected not to appeal against this judgment.

In particular, the judges said that the Skyguide managers had failed to exercise sufficient care by leaving just one air traffic controller in charge of the southern German and eastern Swiss airspace. For years they had tolerated this wrong practice. Therefore, they were not held criminally liable for the behaviour of the controller (indirect liability), but for their own failure to ensure that control workstations were properly staffed at all times (direct liability).

This accident demonstrated the difficulty to ascertain liabilities in a complex technological environment[51] and especially to demonstrate the defectiveness of technology.

This ruling was in a sense inspired by 'just culture' principles, considering that the aim of this culture is to ensure a balanced accountability for both individuals and organizations, responsible for establishing and improving safety systems in the workplace. Organisational deficiencies were recognized in the trial, but criminal liability was laid on the managers of the ANSP as individual employees and not to Skyguide as a legal entity, because corporate criminal liability is not recognised under Swiss Law.

4.1.3   The Gonesse Disaster

Alongside these cases, there are other ones in which 'just culture' principles have been instead properly applied. An emblematic case led to the institution of criminal proceedings in the well-known Gonesse accident, occurred on July 25, 2000.[52] A

---

[48]Chapter 3, Part III, Section 3, para. 3.2 (Use of ACAS indicators).

[49]For a commentary, see Michaelides-Mateou and Mateou (2012), chapter 4; Finocchiaro and Starrantino (2013), p. 32; Johnson (2008).

[50]So Finocchiaro and Starrantino (2013), pp. 31 ff.

[51]See Marais et al. (2004), p. 9; Bruni and Tullio (2013).

[52]See Henley (2001); Byers (2002), pp. 31 ff.; Brookes (2002), p. 22; Woods and Woods (2008), p. 66; Lichfield (2010); Hawkins (2011), p. 32.

Concorde aircraft, operated by Air France (charter flight 4590) from Paris to New York, crashed in Gonesse, a suburb of Paris. The airplane went down in flames almost immediately after take-off, killing all 109 people on board and 4 others on the ground.

The official investigation, carried out by the competent Accident Investigation Bureau (BEA),[53] concluded that, after reaching take-off speed, a wheel tyre of the Concorde involved was cut by a metal strip on the runway lost from a Continental Airlines aircraft[54] a few minutes before the supersonic airplane took-off from the Charles de Gaulle airport in Paris.

Ten years after the accident,[55] a mechanic was found guilty of involuntary manslaughter for this accident.[56]

In fact, during the first instance trial, a French prosecutor filed charges against Continental Airlines and one of its mechanics. Three other people were believed to have been negligent in the disaster: one of them was employed by the company at the time, another was an employee of airplane manufacturer (EADS-France) and the third one was an official at the French civil aviation authority (DGAC).

On 6 December 2010, a French Court decision[57] found Continental Airlines and its mechanic responsible for this disaster,[58] while the other people were all cleared of involvement in the crash. Charges against Air France and the Concorde designers were dropped, and Aéroports de Paris was not held to account over its failure to perform a runway inspection.

In particular the Court determined that the manufacturer[59] EADS-France's responsibility "does not qualify as serious misconduct" but as a less serious offense of negligence.

The Court was convinced that the 16-inch piece of metal (a wear strip) punctured a tire on the jet as it sped down the runway for take-off and that debris perforated the Concorde's low-lying fuel tank, causing a leak and a fire.

---

[53]BEA (Bureau d'Enquête et d'Analyses pour la Sécurité de l'Aviation civile), Initial Inquiry Report, 14 December 2004, Accident on 25 July 2000 at La Patte d'Oie in Gonesse (95) to the Concorde registered F-BTSC, operated by Air France. www.bea.aero/uploads/tx_elydbrapports/f-sc000725a.pdf.

[54]McDonnell Douglas DC-10, operated by Continental Airlines, flight from Paris to Houston.

[55]See Lawrenson and Braithwaite (2018), para. 3.1: "the various prosecutions of individual and corporate manslaughter took over a decade to process through the French legal system", pp. 252 ff.

[56]Rose (2001); Bremner (2008).

[57]Decisions of a courthouse (*Tribunal correctionnel*) in Pontoise, northwest of Paris, 6 December 2010.

[58]Continental was fined about $265,000 for "manslaughter by a legal entity and unintentional injuries". Its mechanic was fined $2650 and ordered to serve a 15-months suspended prison sentence. The Court ruled that Continental should pay Concorde operator Air France 1m euros in compensation.

[59]Three French engineers who headed the manufacturing portion of the supersonic Concorde, including Henri Perrier, the so-called father of the Concorde, were cleared of all charges while their employer, EADS-France, was held civilly liable and ordered to pay 30% of about $250,000 in damages to victims.

A French Appeals Court in Versailles on 29 November 2012[60] overturned the 2010 verdict against Continental, absolving this US carrier of criminal responsibility, without prejudice for civil liability against Air France. It also acquitted the Continental mechanic who had been found guilty of manslaughter. This decision is a reminder that honest human error, regardless of the tragic outcome, is different from a crime. 'Just culture' principles had triumphed in this respect.

During the appeals trial, judicial investigators told that officials had known for years about design problems with the Concorde and that the aircraft should not have been allowed to fly, because of the high risk of accident. Therefore, officials have been charged with reckless endangerment by the Court.

In October 2003, the Concorde fleets were retired[61] by Air France and British Airways, the two airlines that operated this supersonic aircraft.

4.2   The Implications of Criminalization

In light of the aforementioned observations, we shall now analyse the implications of criminalization[62] for aviation safety.

Criminalizing airline accidents suppresses the free flow of safety-related information, with a deleterious effect on aviation safety.

Such a punitive approach thus reduces existing levels of safety in the long term, with consequent negative impact on air transport.[63]

In order to obviate the risks of the punitive approach, a number of States and airlines have recently developed voluntary disclosure programs, which encourage the reporting of safety-related incidents by offering the front-line reporter immunity from disciplinary measures. This is part of a 'just culture' environment.

In conclusion, 'just culture' has been correctly defined as "the growing recognition of the need to establish communication and training initiatives and advance arrangements between the aviation safety sector, regulators, law enforcement and the judiciary to avoid unnecessary interference and to build mutual trust and understanding in the relevance of their respective activities and responsibilities".[64]

---

[60]Decision of Appeals Court in Versailles, west of Paris, 29 November 2012.

[61]The first supersonic passenger-carrying commercial airplane was built jointly by aircraft manufacturers in Great Britain and France. Its first flight was operated in 1969. It made its first transatlantic crossing on September 26, 1973 and inaugurated the world's first scheduled supersonic passenger service on January 21, 1976. In 2003, 3 years after the Gonesse disaster, it was retired. See Cramoisi (2010), pp. 517 ff.; Glancey (2015), para. 10; Chittum (2018), para. 12.

[62]See Licu and van Dam (2013b), p. 19; Licu et al. (2013), pp. 55 ff.; Panelli and Scarabello (2013), pp. 58 ff.

[63]See Pellegrino (2013), p. 480.

[64]So Licu and van Dam (2013a), p. 14; Antonini (2015), pp. 45 ff.; Borzì (2014), p. 73 ff.; Pellegrino (2014); Bufo (2015), pp. 143 ff.

4.3   Justice Versus Aviation Safety with Regard to Safety Investigation and Occurrence Reporting

In general, the main question that underlines the theory of tort liability[65] is how the law must reconcile competing interests.[66]

In aviation, the European legislation over the past few years has stressed that different public interests, such as the prevention of future adverse events and the proper administration of justice, can be achieved by means of accident or serious incident investigation.[67]

Other questions arise: Can justice and safety co-exist?[68] Can judicial investigation and safety investigation exist side by side in a coordinated manner?

An answer to this question can be found in the aviation safety legislation inspired by 'just culture' principles.

These are the public interests, which go beyond the individual interests of the involved parties and the event occurred: on the one hand, the right to the highest level of safety (through accident investigation and occurrence reporting) and, on the other hand, the right of independent and impartial administration of justice, guaranteed by constitutional and legislative provisions, at the international,[69] European[70] and national levels.

Two EU legislative texts in particular, Regulations (EU) No. 996/2010 and No. 376/204, stated that a balance between the above mentioned objectives is required. The right balance among all interests involved is necessary to guarantee the overall public interest.[71]

It is no coincidence that these two texts have been inspired by the modern concept of 'just culture', considering that the first one governs safety investigations and the

---

[65]Epstein (1980); Lawson and Markesinis (1982); Cooper-Stephenson and Gibson (1993); England (1993); Heiderhoff and Żmij (2009), pp. 2 ff.; Beever (2016), p. 28. See McDonald et al. (2002).

[66]So Daniels (2017), chapter 3, para. 3.2

[67]See, in particular, Regulation (EU) No. 996/2010 (recital 23) and Regulation (EU) No. 376/2014 (Article 15(4)).

[68]See Ter Kulle (2004), pp. 1 f.

[69]See ICAO General Assembly Resolution A37-2 e A37-3, mentioned above.

[70]For the right to safety at European level, see Article 6 of the Charter of Fundamental Rights of the European Union (Nice, 7 December 2000), entitled "Right to liberty and security" ("Everyone has the right to liberty and security of person"). We remember that, according to Article 6 of the Treaty on the Functioning of the European Union (TFEU), "the *Union* recognises the rights, freedoms and principles set out in the Charter of Fundamental Rights of the European Union of 7 December 2000, as adapted at Strasbourg, on 12 December 2007, which shall have the same legal value as the Treaties". The Charter, therefore, constitutes primary EU law; as such, it serves as a parameter for examining the validity of secondary EU legislation and national measures. It is a legally binding instrument that was drawn up in order to expressly recognise, and give visibility to, the role of fundamental rights in the legal order of the Union. So European Parliament (ed), Respect for fundamental rights in the European Union, Fact Sheets on the European Union, 16/04/2018.

[71]So recital 23 of Regulation (EU) No. 996/2010, frequently mentioned.

second one regulates occurrence reporting. In fact, there is a strict relationship between 'just culture', safety investigation and reporting of accidents and incidents.[72] The relationship between safety investigation/reporting and 'just culture' is comparable to the link existing between a means and an end. In fact, a process for reporting, tracking, and investigating allows to achieve a 'just culture'.

Before the adoption of the already mentioned Council Directive 94/56/CE, any balance between these different interests was missing with regard to accident technical investigation. In fact, in many EU countries justice requirements was intended to take precedence and to prevail over safety requirements. In situations where both a safety investigation and a judicial inquiry had been opened, the first one was obstructed by the seizure of the aircraft or its wreckage and any evidence (the accident area included), ordered by the competent judicial authority.

On the contrary, this Directive requested all Member States, in the framework of their respective internal legal systems, to define a status for safety investigation[73] that allowed the investigators-in-charge 'to carry out their task in the most efficient way and within the shortest time' (Article 5).

The rights of the safety investigator included the following: to have free access to the site of the accident or incident as well as to the aircraft, its contents or its wreckage; to ensure an immediate listing of evidence and controlled removal of debris or components for examination or analysis purposes; to have immediate access to and use of the contents of the flight recorders and any other recordings; to have access to the results of examination of the bodies of victims or tests made on samples taken from the bodies of victims; to examine witnesses; to have free access to any relevant information or records held by the owner, the operator or the manufacturer of the aircraft and by the authorities responsible for civil aviation or airport operation.[74]

Once national legislation was sufficiently harmonized, the European legislator adopted a new legislative act, entirety and directly applicable in all Member States, the well-known Regulation (EU) No. 996/2010, examined above, that repealed Directive 94/54/CE. However, according to Article 11 of this Regulation, "the investigator-in-charge shall have the authority to take the necessary measures to satisfy the requirements of the safety investigation". Therefore, the safety investigator has the same rights and powers already laid down in the repealed Directive.

The main objective of this Regulation is to allow technical investigations to be conducted diligently and efficiently.[75]

In this context, a very important provision is contained in Article 12 that promotes strict coordination between both safety and judicial investigation. Despite their full autonomy, the first one is aimed to ensure accident prevention, while the second one is aimed to apportion blame and responsibility.

---

[72]See above, chapter II, para. 2.1. See Masys (2016a).

[73]Comenale Pinto (2015), pp. 237 ff.

[74]So Article 5 (2), letters a)–g), of the Council Directive 94/56/CE.

[75]See Article 12(3) of Regulation (EU) No. 996/2010.

According to this Article, when both safety and judicial investigation are opened at the same time for the same accident, the investigator-in-charge shall guarantee protection of flight recorders and any other physical evidence. The judicial authority may appoint an official to accompany the flight recorders or physical evidence to the place of the read-out or treatment. Where examination or analysis of such evidence may modify, alter or destroy it, prior agreement from the judicial authorities will be required, without prejudice to national law. Where such advance agreement on non-repeatable inspection is not obtained within a reasonable time and in any case not later than 2 weeks following the request, the investigator-in-charge shall be allowed to conduct the examination or analysis. However, where the judicial author-ity is entitled to seize any evidence, the investigator-in-charge shall have immediate and unlimited access to and use of such evidence.

Article 12(3), states that Member States shall ensure that safety investigation authorities and other authorities involved in the activities related to the safety investigation (such as the judicial, civil aviation, search and rescue authorities), cooperate with each other through advance arrangements. These arrangements shall respect the independence of the safety investigation authority.

Pursuant to the latter provision, the Italian investigation authority (ANSV) and the Ministry of Justice prepared a draft of advance agreement. This agreement, that was signed by only a few public prosecutor's offices,[76] covers the following aspects: preservation of the site concerned, arrangements for access to the crash site, collec-tion and preservation of impounded evidence and access to it, acquisition of the data held in flight recorders, the carrying out of non-repeatable expert inspections and autopsy scans, resolution of conflicts in order to implement the advance agreement.

The above mentioned provisions have been examined to demonstrate the inten-tion of the European legislature to achieve a balance between a safety investigation, whose only purpose is prevention of accidents and incidents,[77] and a judicial investigation, aimed to apportion blame or liability.

In the field of occurrence reporting for civil aviation we can find the same legislative evolution towards a balance between safety and justice.[78]

In fact, whereas Directive 2003/42/EC did not contain any reference to the need to balance and reconcile these different interests, the new Regulation (EU) No. 376/2014 amply provides for it.

---

[76]A letter of formal notice has been notified by the European Commission to the Italian Republic according to Article 258 TFEU (on the infringements procedure), because the Commission has not yet received all prior agreements, signed by all (153) public prosecutor's offices, under Article 12 (3), Regulation (EU) No. 996/2010.

[77]So Article 2(14) of Regulation (EU) No. 996/2010, frequently mentioned, that contains the following definition of "safety investigation": 'safety investigation' means a process conducted by a safety investigation authority for the purpose of accident and incident prevention which includes the gathering and analysis of information, the drawing of conclusions, including the determination of cause(s) and/or contributing factors and, when appropriate, the making of safety recommendations'.

[78]Ter Kulle (2004), pp. 1 ff.

According to recital 45, the cooperation between safety authorities and judicial authorities should be 'enhanced and formalised by means of advance arrangements between themselves', with a view to balancing the different public interests. This cooperation, in particular, should cover access to, and the use of, occurrence reports contained in the national databases.

Consequently, Article 15(4) states that Member States shall ensure that their competent authorities designated to establish a mechanism to independently collect, evaluate, process, analyse and store details of occurrences reported and their competent authorities for the administration of justice cooperate with each other through advance administrative arrangements.

These arrangements shall seek to ensure the correct balance between the need for proper administration of justice, on the one hand, and the necessary continued availability of safety information, on the other hand.

References

Antonini A (2015) La responsabilità nell'esercizio del volo. In: Pellegrino F (ed) Legislation and regulation of risk management in aviation activity, vol II. Giuffré, Milano, pp 45–54

Barra B (2013) La giurisprudenza dell'ultimo decennio: prevalenza della norma penale sulle regole tecniche internazionali e comunitarie; impatto su compiti e responsabilità degli operatori aeronautici. In: Bruni F, Tullio L (eds) Sinistri aeronautici: rischi e responsabilità. Giappichelli, Torino, pp 8–27

Bate P (1992) The impact of organisational culture on approaches to organisational problem-solving. In: Salaman G (ed) Human resource strategies. Sage, London, pp 229 ff

Beever A (2016) A theory of tort liability. Bloomsbury, Oxford and Portland

Bennet S (2004) The 1st July 2002 mid-air collision over Überlingen, Germany: a holistic analysis. Risk Manag 6(1):31–49

Blajev T (2013) Just culture in doubt. HindSight 18:6–7

Borzì I (2014) Just culture and responsibility of air traffic controllers: some considerations regarding Judgement No. 6820/11 of the Italian criminal Court of Cassation. In: Pellegrino F (ed) Legislation and regulation of risk management in aviation activity, vol I, pp 73–98

Bremner C (2008) Continental Airlines faces manslaughter charges over Paris Concorde crash. The Times, London

Brookes A (2002) Destination disaster. Ian Allan, Birmingham, pp 22 ff

Bruni F, Tullio L (eds) (2013) Sinistri aeronautici: rischi e responsabilità. Giappichelli, Torino

Bufo M (2015) Gestione del rischio d'impatto tra aeromomili e volatile (c.d. Bird strike) in aeroporto e nello spazio aereo limitrofo: ripartizione delle competenze (e responsabilità) tra i principali soggetti coinvolti. In: Pellegrino F (ed) Legislation and regulation of risk management in aviation activity, vol II. Giuffré, Milano, pp 123–154

Byers A (2002) The crash of the Concorde. The Rosen Publishing, New York

Camarda G (2002) La responsabilità del gestore aeroportuale. Diritto dei trasporti 3:763–811

Cattaneo C, De Angelis D, Grandi M (2006) Mass disasters. In: Schmitt A, Cuhna E, Pinheiro J (eds) Forensic anthropology and medicine. Complementary sciences from recovery to cause of death. Humana Press, Totowa, pp 431–443

Chittum S (2018) Last days of the Concorde: the crash of Flight 4590 and the end supersonic passenger travel. Smithsonian Books, Washington

Coelho dos Santos J (2013) Just culture versus criminalization – moving forward. HindSight 18:40 ff

Comenale Pinto MM (2015) Inchieste aeronautiche, raccomandazioni di sicurezza e "just culture". Diritto marittimo, pp 237 ff

Cooper LC, Robertson I (eds) (1988) International review of industrial and organisational psychology. Wiley, Chichester

Cooper-Stephenson K, Gibson E (eds) (1993) Tort theory. Captus University Publications, North York

Cramoisi G (2010) Air crash investigations: the end of the Concorde era. The crash of Air France Flight 4590. Mabuhay Publishing, New York

Croucher P (2015) Avionics in plain English. Electrocution, Calgary, pp 2–49

Daniels S (2017) Corporate manslaughter in the maritime and aviation industries. Informa law. Routledge, Taylor and Francis Group, Abingdon

De Meo MM (2000) Riassetto del comparto aeroportuale in Italia. Gazzetta ambiente, pp 3 ff

Deiana M (2006) La definizione di disastro aviatorio nell'ordinamento penale. Diritto dei trasporti, pp 52 ff

Dekker S (2013a) A new just culture algorithm. HindSight 18:8–9

Dekker S (2013b) Sicurezza e pensiero sistemico. La gestione della sicurezza nelle organizzazioni di oggi richiede di passare dalla caccia alle componenti rotte alla comprensione dei sistemi complessi. Hirelia ed., Milano

Dekker S, Laursen T (2007) From punitive action to confidential reporting. Patient Saf Qual Healthc 5:50–56

Del Pinto G (2017) Il fattore umano nel controllo del traffico aereo. I quaderni di ANACNA, vol 1. Youcanprint, Roma

DeWinter D (2013) Experience on task in a just culture. HindSight 18:38–39

Di Palma P (2000) Il ruolo dell'ENAC per l'indirizzo, la regolazione e il controllo del settore. Gazzetta ambiente 2–3:11–12

Dusi G (2017) Just culture safety report: atteggiamenti e comportamenti dei professionist. Youcanprint Self-Publishing, Tricase

Englard I (1993) The philosophy of tort law. Dartmouth Publishing Company, London

Epstein RA (1980) A theory of strict liability: toward a reformulation of tort law. Cato Institute, Washington

Fassert C (2001) Questioning transparency: the case of incidents. MSc thesis (mémoire de DEA). http://www.eurocontrol.int

Fassert C (2005) A comparative approach on safety in several air navigation services providers: which role for culture? ATRS (Air Transport Research Society International Conference), Rio de Janeiro

Finocchiaro M, Starrantino C (2013) The Ueberlingen case: legal scenarios after Barcelona Court of Appeal judgment. The Controller Magazine 51(4):31–33

Franchi B (2011) Controllori del traffico aereo: un'altra sentenza che ne amplia le competenze. Responsabilità civile e previdenza 11:2284–2295

Garland DJ, Wise JA, Hopkin VD (eds) (1999) Handbook of aviation human factors. Lawrence Erlbaum Associates Publishers, Mahwah

Garzone G, Archibald J (eds) (2010) Discourse, identities and roles in specialized communication. Peter Lang, Bern

Gates S (2009) Criminalisation of air accidents – time to turn the tide. Reg Int 53:9 ff

Glancey J (2015) Concorde: the rise and fall of the supersonic airliner. Atlantic Books, London

Hawkins J (2011) Air disasters. The Rosen Publishing, New York

Hayward BJ, Lowe AR (eds) (2000) Aviation resource management. Ashgate Publishing Limited, Aldershot

Heiderhoff B, Żmij G (eds) (2009) Tort law in Poland, Germany and Europe. Sellier, European Law Publishers, Munich

Hempel Lindøe P, Baram M, Renn O (eds) (2014) Risk governance of offshore oil and gas operations. Cambridge University Press, Cambridge

Henley J (2001) Concorde crash 'a disaster waiting to happen'. The Guardian, 17 August 2000

Hofstede G (1991) Cultures and organisational: intercultural co-operation and its importance for survival. Harper Collins, London

Johnson CW (2008) Have we learned enough from Überlingen: the challenges of safety improvement in European Air Traffic Management, University of Glasgow, Glasgow. https://pdfs. semanticscholar.org/5856/b38ce8b127ee974f19ec6f7b8415c2cc67db.pdf and http://www.dcs. gla.ac.uk/~johnson/

Kaminski-Morrow D, Learmount D (2004) Spate of errors led to Linate crash. Flight International, London

Kodate N, Dodds A (2008) Factors affecting willingness to report patient safety incidents in hospitals. NIHR King's Patient Safety and Service Quality Centre (PSSQ) Risk Programme, Working paper 1, Cambridge University Press, Cambridge, pp 2–38

Lawrenson AJ, Braithwaite GR (2018) Regulation or criminalisation: what determines legal standards of safety culture in commercial aviation. Safety Sci 102:251–262

Lawson FH, Markesinis BF (1982) Tortious liability for unintentional harm in the common law and the civil law, vol I. Cambridge University Press, Cambridge

Lichfield J (2010) Air France grounds Concorde until cause of crash is known. The Independent, London, 18 October 2010

Licu T, van Dam R (2013a) Just culture in aviation: dynamics and deliverables. In: Pellegrino F (ed) Legislation and regulation of risk management in aviation activity, vol I. Giuffrè, Milano, pp 55–65 ff

Licu T, van Dam R (2013b) Just culture in aviation: dynamics and deliverables. HindSight 18 (2013):18–21

Licu T, Baumgartner M, van Dam R (2013) Everything you always wanted to know about just culture (but we afraid to ask). HindSight 18:14–17

Magnosi S (2010) Il nuovo assetto organizzativo dell'Ente Nazionale per l'Aviazione Civile. Rivista di Diritto della navigazione, pp 313–318

Marais K, Dulac N, Leveson N (2004) Beyond normal accidents and high reliability organizations: the need for an alternative approach to safety in complex systems. Engineering Systems Division Symposium, MIT, Cambridge, pp 1–16

Martorell S, Guedes Soares C, Barnett J (eds) (2014) Risk analysis: theory, methods and applications. CRC Press, Taylor and Francis, Boca Raton

Marx D (2012) Just Culture Algorithm™ V3. 2 – for employers. The Just Culture community. Outcome Engenuity, Eden Prairie

Masutti A (2010) Prospettive di riforma dell'aviazione civile. Diritto dei trasporti, pp 473–483

Masys A (ed) (2016a) Disaster forensics: understanding root cause and complex causality. Springer, Ottawa

Masys A (2016b) Patient safety and disaster forensics: understanding complex causality through actor network ethnography. In: Masys A (ed) Disaster forensics: understanding root cause and complex causality. Springer, Ottawa, pp 63–82

McDonald N, Corrigan S, Cromie S, Daly C (2000) An organizational approach to human factors. In: Hayward BJ, Lowe AR (eds) Aviation resource management. Ashgate Publishing Limited, Aldershot, pp 51–61

McDonald N, Corrigan S, Ward M (2002) Well-intentioned people in dysfunctional systems, Keynote presented at Fifth Workshop on Human Error, Safety and Systems Development, Newcastle

Michaelides-Mateou S, Mateou A (2012) Flying in the face of criminalization: the safety implications of prosecuting aviation professionals for accidents. Routledge, Abingdon-on-Thames

Murray SR (1997) Deliberate decision making by aircraft pilots: a simple reminder to avoid decision making under panic. Int J Aviat Psychol 7(1):83–100

Nemsick J, Passeri SG (2012) Criminalizing aviation: placing blame before safety. American Bar Association Journal, pp 21 ff

Orasanu JM (1993) Decision-making in the cockpit. In: Weiner EL, Kanki BG, Helmreich RL (eds) Cockpit resource management. Academic, San Diego, pp 137–172

Panelli S, Scarabello M (2013) Why is it necessary to criminalise negligent behaviour? HindSight 18:58–61

Pellegrino F (2013) Just culture principles in aviation law from a European perspective. Ann Air Space Law, pp 471–490

Pellegrino F (ed) (2014) Legislation and regulation of risk management in aviation activity, vol I. Giuffré, Milano

Reason J (1995) Comprehensive error management in aircraft engineering: a manager's guide. British Airways Engineering, London

Rose D (2001) Concorde: the unanswered question. The Guardian, 13 May 2001

Rousseau D (1988) Quantitative Assessment of Organisational Culture: the case for multiple measures. In: Cooper LC, Robertson I (eds) International review of industrial and organisational psychology. Wiley, Chichester, pp 139–158

Salaman G (ed) (1992) Human resource strategies. Sage, London

Schein EH (2010) Organizational culture and leadership, IV edn. Jossey-Bass, San Francisco

Schmitt A, Cuhna E, Pinheiro J (eds) (2006) Forensic anthropology and medicine. Complementary sciences from recovery to cause of death. Humana Press, Totowa

Shorrock S (2013) 'Human error'. The handicap of human factors. HindSight 18:32–37

Simpson P (2014) The Mammoth Book of air disasters and near misses. Running Press Book Publishers, Philadelphia

Sirena PA (2014) The case law of the Italian Supreme Court of Cassation in the field of aviation accident. Eurocontrol Seminar on "Aviation and the Judiciary", 30 October 2014, Bruxelles

Ter Kulle A (2004) Safety versus justice. CANSO News, 18, pp 1–2

Thompson N, Stradling S, Murphy M, O'Neil P (1996) Stress and organizational culture. Br J Soc Work 26(5):647–665

Trögeler M (2010) Criminalisation of aviation accidents: creation of a just culture: balancing aviation safety and the proper administration of justice. Leiden University, Law School, International Institute of Air and Space Law, Leiden

Trögeler M (2011) Criminalisation of air accidents and the creation of a Just Culture. Diritto dei trasporti, pp 1–44

Turney R (2007) The Überlingen mid-air collision: lessons for the management of control rooms in the process industries. Institution of Chemical Engineers (IChemE) Symposium Series, 153, pp 1–5

Weick KE (1987) Organizational culture as a source of high reliability. Calif Manag Rev 29:112–127

Weick KE, Roberts KH (1993) Collective mind in organizations: heedful interrelating on flight decks. Adm Sci Q 38(3):357–381

Weiner EL, Kanki BG, Helmreich RL (eds) (1993) Cockpit resource management. Academic, New York

Westrum R, Adamski AJ (1999) Organizational factors associated with safety and mission success in aviation environments. In: Garland DJ, Wise JA, Hopkin VD (eds) Handbook of aviation human factors. Lawrence Erlbaum Associates Publishers, Mahwah, pp 67–104

Woods M, Woods MB (2008) Air disasters. Lerner, Minneapolis

Chapter 5
From a 'Blame Culture' to a 'Just Culture' and Back: The Italian Experience in the Aviation Field

5.1 An Italian Trial Inspired by 'Blame Culture' Principles: Aftermath of the "Monte dei Sette Fratelli" Disaster

On 24 February 2004, a Cessna 500 Citation I, operated by the Austrian air taxi operator City-jet (CIT flight number 124), inbound to Cagliari airport, approved for a visual approach, collided against the rock, so-called 'Baccu Malu', that is part of the 'Monte dei Sette Fratelli'. The flight had to transport a medical team composed of three doctors and an organ for transplantation. All six occupants (three crew members and three medical team members) died in the crash.

In the judgment of 10 December 2010,[1] concerning the 'Monte Sette Fratelli' crash,[2] which occurred in 2004, the Italian Supreme Court of Cassation sentenced

---

[1]Italian Supreme Court of Cassation, criminal division, IV section, No. 2019 of 10 December 2010-22 February 2011.

[2]It may be interesting provide a few air-ground voice conversations. At 5.41 Cagliari APP (Approach Control Service) instructed CIT 124 (flight number), which at that time was at the waypoint Aledo, to continue the descent over the next waypoint Ledro up to 5000 ft, to carry out the ILS procedure for a landing on runway 32 and report at Carbonara VOR (Very High Frequency Omnidirectional Range). The commander, reading back the instructions, requested permission to conduct a visual approach as follows: "we are cleared now to descend to five thousand feet on the QNH one zero one two and we passing Ledro cleared by ILS three two papa and if we have fied visually, we would like to have a visual approach". This request was confirmed by Cagliari APP. Then the flight crew radioed: "CIT 124, we have the field in sight requesting visual approach". At the time the aircraft was at FL 96, about 28 miles from Carbonara VOR. Cagliari APP then asked CIT 124 about the ability to maintain obstacle clearance: "confirm able to maintain your own separation from the obstacles sir, performing visual approach runway three two". The pilot just replied: "affirmative, CIT 124". Three minutes later, Cagliari APP informed CIT 124 to continue the descent without descending below 2500 ft and contact Elmas TWR for the further descent. The pilot contacted Elmas TWR: "Elmas Tower, buongiorno, CIT 124, visual, left procedure, for visual approach runway 32". Elmas TWR responded by reporting the runway in use (32), wind conditions (8 knots from 310°) and told the flight to report on short final for runway 32. This was read back by

© Springer Nature Switzerland AG 2019
F. Pellegrino, *The Just Culture Principles in Aviation Law*, Legal Studies in International, European and Comparative Criminal Law 3,
https://doi.org/10.1007/978-3-030-23178-1_5

the two military air traffic controllers involved in the criminal proceeding for giving a night visual approach[3] clearance[4] "without supporting the pilot in providing all the necessary information on terrain" where the aircraft was assumed to land. In its view, their negligent conduct had made a substantial contribution to the fatal event.

In Annex 13 and other ICAO documents, we can find the difference between 'causes' and 'contributing factors'. The first ones are intended as "actions, omissions, events, conditions, or a combination thereof, which led to the accident or incident".[5] The second ones are intended as "actions, omissions, events, conditions, or a combination thereof, which, if eliminated, avoided or absent, would have reduced the probability of the accident or incident occurring, or mitigated the severity of the consequences of the accident or incident".[6]

During the Cagliari crash process, technical experts appointed by the judicial authority[7] concluded that the main cause of the disaster was the decision of the commander

to execute a visual approach in a situation that did not allow to maintain an appropriate obstacle clearance both in order to the characteristics of the terrain of this geographical area and to the absence of luminous visual references that inhibited, during the night and in those conditions, to see obstacles and to be separated from them.

---

the crew. Two minutes later, while descending over mountainous terrain, flight 124 collided with the Baccu Malu Mountain (3333 ft, speed 226 knots). Source: Aviation Safety Network, an exclusive service of Flight Safety Foundation, website https://aviation-safety.net. According to the reenactment of investigators, the accident, classified as CFIT (controlled flight into terrain), was caused by the conduct of the flight at a height significantly below the Area inimum Altitude, insufficient to maintain the separation from the ground, during a night visual approach, in absence of adequate visual reference. For a technical reconstruction of the dynamic of the crash see Pooley (2013), p. 71 ff. See also 'Incidente di Cagliari 2004: la storia complete', in www.anacna.it/anacna-news/documenti/legal/170-incidente-di-cagliari-2004-la-storia-completa.html.

[3]A visual approach is defined as "An approach by an IFR flight when either part or all of an instrument approach procedure is not completed and the approach is executed in visual reference to terrain". Instrument approach procedure (IAP) is defined as "A series of predetermined manoeuvres by reference to flight instruments with specified protection from obstacles from the initial approach fix, or where applicable, from the beginning of a defined arrival route to a point from which a landing can be completed and thereafter, if a landing is not completed, to a position at which holding or en-route obstacle clearance criteria apply" (ICAO Doc 4444, chapter 1, see below). The same document, at point 6.5.3 (Visual approach) specifies the conditions the requirements to be satisfied for a visual approach.

[4]For a commentary of this ruling, see Bufo (2012), p. 163 ff.; Starrantino and Finocchiaro (2013), p. 74 ff.; Scala (2013), p. 80 f.; Pellegrino (2013), p. 482, Pellegrino (2014), Pellegrino (2015); Sirena (2015b), p. 6 ff.; Sirena (2013); Barra (2013a), p. 8 ff.; La Torre (2015a, b), p. 2 ff.

[5]See Annex 13, chapter 1; Jing and Battea (2016), chapter 2.

[6]See ICAO Model Accident Investigation Authority Act, second edition, March 2017. See Bruni and Tullio (2013).

[7]See Pooley (2013), p. 71 ff. This paper summarises the findings, conclusions and recommendation of the ANSV (Agenzia Nazionale Sicurezza del Volo) investigation, carried out under ICAO Annex 13 principles. It is based on the Final Report of the Agency, which was published on 1 July 2009.

However, even though the experts identified the wrong pilot decision as the main cause of this controlled flight into terrain (CFIT),[8] the Supreme Court wondered how the pilot was allowed to dangerously descend undisturbed at this low altitude, without any intervention by the air traffic controllers.

Thus, despite the conclusions of judicial experts who testified that the two controllers involved had followed the technical rules, regulations and procedures in force, both at the international (ICAO Annex 11 and Doc 4444) and national levels (AIP[9] and AMI-Air Force ATS Manual[10]), they were prosecuted and sentenced.

Nevertheless, the verdicts of the Italian Tribunal of Cagliari[11] and the Court of Appeal,[12] confirmed by the Supreme Court of Cassation, condemned the two ACTs for multiple manslaughter (Article 589 of the Italian Criminal Code)[13] and culpable aviation disaster (Article 449 of the same Code),[14] because their behaviour has been considered culpable and treated as a contributory cause of the accident.

In particular, Article 449[15] contemplates the crime of culpable aviation disaster. This is the case of an accident that endangers public safety. Under Italian Criminal Law, if a crash causes death and/or injuries and also endangers public safety (for example, when the aircraft was flying over populated areas), both manslaughter/ culpable injuries and culpable aviation disaster are alleged at the same time.

---

[8]A controlled flight into terrain (CFIT) is an accident in which an airworthy aircraft under pilot control, is unintentionally flown into the ground, a mountain, a waterbody (ocean, sea, lake, river etc.) or an obstacle. See Smith (2001).

[9]AIP is the acronym derived from Aeronautical Information Publication. It is defined by ICAO Annex 15 (Aeronautical Information Services), Fifteenth Edition, July 2016) as "a publication issued by or with the authority of a State and containing aeronautical information of a lasting character essential to air navigation" (Article 1(1)). AIP Italia (Servizio Informazioni Aeronautiche), published by ENAV (Italian Company for Air Navigation Service) contains all permanent aeronautical information concerning the national airspace, the airports, the organization of air traffic services, infrastructures etc. The last edition, dated 22 December 2016, No. 13/16, entered into force on 2 February 2017.

[10]AMI (Aeronautica Militare Italiana) - ATS (Air Traffic Services) Manual is published by the Italian Air Force.

[11]Judgment of the Tribunal of Cagliari, criminal division, of 17 March 2008, that condamned the two ATCs to 3 years' imprisonment (reduced to 2 years due to summary judgment chosen) and to pay, jointly, an interim compensation amount of 75,000 € for civil liability and justice costs. The conviction was based on the fact that the controllers allowed the visual approach "without supporting the pilot in providing all the necessary information on terrain", as required, according to the judges view, by additional conditions on the night visual approach, issued by the then Directorate General of Civil Aviation (DGAG) of the Italian Ministry of Transport. For a commentary see Barra (2013b); Starrantino and Finocchiaro (2013), p. 74 f.; Sirena (2015a, b), p. 252 ff.

[12]Judgment of the Court of Appeal of Cagliari, criminal division, 18 March 2010. See Barra (2013b); Sirena (2015a, b), p. 252 ff.

[13]For a commentary, see *infra*.

[14]Punishable with a sentence of 2–10 years imprisonment.

[15]The crime of culpable aviation disaster is punishable with a sentence of 2–10 years imprisonment. See Alicke (2000), p. 557 ff.

In Italian law, in which criminal prosecution is allowed not only in case of accident, but also in the event of an incident characterised by a relevant risk to public safety, in the latter case the existence of a real danger must often be demonstrated and requires an extensive technical and practical experience of the aviation environment reality.

In the present case, the Courts departed from the results of the judicial inquiry by observing that the experts just verified that the controllers had complied with the specific visual approach operational requirements, published in AIP Italia, and in turn contained in Doc 4444 (so-called ICAO PANS ATM, mentioned above), that obliged them to separate aircrafts from each other, but not from obstacles.[16]

In fact, RAC 1-47, part. IV, paragraph 9.1.1, of AIP Italia,[17] that refers to the ICAO Doc 4444, says "Separation shall be provided between an aircraft cleared to execute a visual approach and other arriving and departing aircrafts".[18]

This provision is very important because it implies that separation from terrain and obstacles is the responsibility of the pilot.

Instead, according to the judges' opinion, the controllers, in case of a night visual approach, were required to comply with additional operating requirements and restrictions (in the phases of planning and conduct of the flight), contained in the notes No. 41/8879 and No. 41/8880/A.M.O[19] from the Civil Aviation General Direction (DGAC)[20] of the Italian Ministry of Transport, dated 21 June 1991 and implemented by ENAV's decision No. 264 of 7 November 1996.[21]

Since the first instance judgement of the 2004 Cagliari disaster, the application to the air traffic controllers of these notes has been a very controversial issue.

In particular, it is the violation of the requirements imposed by the note No. 41/8880 that was contested, since the controllers did not verify that the pilots

---

[16]Iovino (2013), p. 78 ff.

[17]AIRAC dated 2 October 2003 (A7/03), Article 9(1) (Visual Approach). In the report of the disaster, issued by the ANSV (in http://www.ansv.it/it/detail_Relazioni.asp?ID=1134), it was stressed that AIP Italia which was into force at the time of the accident, referred to the Doc 4444, part IV, p. 9, making reference to an outdated edition of this document, evidently. The edition in force at the time of the accident was the 14th edition, of 2001, not longer split in different parts.

[18]On this subject, see La Torre (2016), p. 75 ff.; Gallo (2002). See also Oxford Aviation Academy (2008), p. 145; Gibb et al. (2016), p. 4; Federal Aviation Administration (2017).

[19]The first note (No 41/8879) on operating conditions specifically prohibited to use *visual approach* procedures at night (from 30 min after sunset, to 30 min before sunrise) but only for general aviation traffic. On the contrary, it removed the ban of night visual approach for commercial air transport (CAT) and aerial work, already introduced on 17 April 1991. But the case was not within the prohibition, considering that the flight CIT 124 formed part of commercial air transport operations. The second one (41/8880) required six operating pre-conditions to be satisfied by flight admitted to operate a *night visual approach:* participation in training course on night visual flight; the knowledge of the terrain around the destination airport; the unavailability to use an alternative instrument approach procedure; an efficient transponder in Mode C and a radio altimeter; visibility of 5 miles and *ceiling* higher than that at beginning of the procedure. See Hunter and Baker (2000).

[20]Now this authority has been merged into ENAC, the Italian Civil Aviation Authority.

[21]For a commentary see, in particular, Battiati (2011), p. 656.

knew the terrain surrounding the Cagliari Elmas airport and did not provide them with relevant information in this respect.

According to the latter note, the pilots of aircrafts operated by public transport airlines (for both passengers and goods) shall be cleared for night visual approach by the controller only under the following conditions, additional to those required by Doc 4444-RAC/501/12[22]:

> a) they shall have completed a training course on the phenomena linked to a night visual approach, with particular reference to visual illusion[23]; b) they must have received information about the specific type of approach procedure, taking into account the specific terrain in the destination aerodrome surroundings (according to the needs, through video tapes, slides, reconnaissance, briefing, information on the approach chart, and all news about obstacles, terrain and potential visual illusion); c) it must be impossible to use instrumental procedure; d) a mode C transponder and a working radio altimeter (set for a prominent warning to at least 500 ft above ground level) must be on board the aircraft; e) at least three miles of vision are required, the ceiling[24] must be set above the start-of-procedure level and constant visual contact with the terrain must be maintained.[25]

It should be stressed,[26] however, that the note No. 41/8880, recalled by the AIP Italia, was not applicable to the controllers.

This is also demonstrated by the fact that the AIP in force at the time of the accident did not mention Article 6(5.3.2)[27] of ICAO Doc 4444, referred to the ATCs, which reads as follows: "Controllers shall exercise caution in initiating a visual approach when there is reason to believe that the flight crew concerned is not familiar with the aerodrome and its surrounding terrain".[28]

In fact, as ascertained by the court-appointed experts since the first instance process, these DGAC's notes had been only addressed to the Administration of the Airport Districts (DD.AA.CC.), with a view to being subsequently disseminated

---

[22]PANS-RAC, Doc ICAO 4444-RAC/501/12, *Procedures for Air Navigation Services*, "Rules of the Air and Air Traffic Services", 12th edition, Montreal, 1985.

[23]This phenomenon is the so-called "black hole approach", a visual illusion that exists on dark nights (usually with no moon or starlight), when there are no ground lights between your aircraft and the runway threshold.

[24]A 'ceiling' is the maximum density altitude an aircraft can reach under a set of conditions.

[25]AAAVTAG (in short, ANAV) is the acronym for Azienda Autonoma Assistenza al Volo per il Traffico Aereo Generale (Italian Agency for Air Navigation Services), which was created in 1981. See Pugliese (1989), p. 232; Serrao (1989), p. 863. This acronym is not longer used because the transformation, in 1996, of the AAAVTAG in ENAV (Ente Nazionale per l'Assistenza al Volo), the Italian company for air navigation services. It was a public commercial institution. See Romanelli (1994a, b), p. 5 ff.; La Penna (1992), p. 182; La Penna (1993), p. 379; Freni (1997), p. 605. From 1 January 2001 ENAV was transformed into a public limited company. See also Comenale Pinto (1997), p. 355; Comenale Pinto (1999) pp. 104 ff. and 245 ff.; Comenale Pinto (2007), p. 107 ff.; Franchi and Vernizzi (2007); Grigoli (1997), p. 23 ff.

[26]So the ANSV Report, cited above, p. 41 f.

[27]This subparagraph is part of the paragraph 6.5.3 (Visual approach), ICAO Doc 4444, cit.

[28]The para. 6.5.3.2 goes on to say: "Controllers should also take into consideration the prevailing traffic and meteorological conditions when initiating visual approaches".

among all airlines operating aircraft engaged in the public carriage of passengers or goods.

Therefore, the experts stressed that these requirements, additional to the provisions of ICAO PANS 4444, were applicable only to pilots and were not binding for air traffic controllers operating in Italy.

According to the experts' opinion, in case of visual approach flight, the relationship between pilots and controllers is based on a statement of the pilot in command. The controller on duty does not have the power to verify the correctness of the visual approach procedures and, where appropriate, to apply sanctions.

This position was in line with that of the Italian Aircraft Accident Investigation Agency (ANSV). In the report[29] concerning the Cagliari disaster, the investigator-in-charge argued that 'the regulatory national framework is not clear and it adds up to an inaccurate information dissemination system (AIP Italia), that is not in line with legislative developments on this subject'.

In fact, the note No. 41/8880 had been overtaken then by the DGAC Circular 41/23100/M3, entitled 'Operating rules for the operation of the aircraft for public transport service',[30] which, as regards visual approach, only specified that 'An operator shall not use an RVR[31] of less than 800 m for a visual approach'.

Subsequently, Decree No. 38-T of the Italian Ministry of Transport, dated 30 March 1998, laying down the adoption of European rules for technical and operational management of commercial air transport aircraft, transposed the JAR-OPS 1[32] into the Italian legal system.

---

[29]See "Relazione d'inchiesta. Incidente occorso all'aeromobile Cessna 500 Citation, marche OE-FAN Punta Su Baccu Malu, Comune di Sinnai (Cagliari), 24 febbraio 2004". In http://www.ansv.it/cgi-bin/ita/ANSV%20OE-FAN.pdf.

[30]Edition January 1997. This document, in the preamble, stated that its provisions "shall annul any other provision on this subject in contrast to these: they also respond to the requirements of JAR OPS-1 Subpart E (All Weather Operations). Operating Licence holders for public air transport service shall be subject to them".

[31]RVR (Runway visual range) is, in aviation meteorology, the distance over which a pilot of an aircraft on the centreline of the runway can see the runway surface markings delineating the runway or identifying its centre line.

[32]JAR-OPS, part 1, is the Joint Aviation Requirement for the operation of commercial air transport (aeroplanes). It was applicable to operators of aircraft over 10 tonnes MTOW (max take-off weight) or with a maximum approved passenger seating configuration of 20 or more, or with mixed fleets of aeroplanes above and below this threshold. So https://www.skybrary.aero/index.php/EU-OPS. JAR-OPS 1 has been transposed into Community Law by the Regulation (EC) No. 1899/2006 of the European Parliament and of the Council of 12 December 2006 amending Council Regulation (EEC) No. 3922/91 on the harmonisation of technical requirements and administrative procedures in the field of civil aviation (O.J. L 377 of 27 December 2006). The JAR (Joint Aviation Requirements) were a set of common comprehensive and detailed aviation requirement issued by the Joint Aviation Authorities, an associated body of the ECAC (European Civil Aviation Conference) representing the civil aviation regulatory authorities of a number of European States who had agreed to co-operate in developing and implementing common safety regulatory standards and procedures. For more details, see Cecchi (1993), p. 129 ff.; Franchi (1993); Arrigoni (1992), p. 130 ff.; Pellegrino (2007), p. 217; Pellegrino (2012a, b), p. 31.

Nevertheless, there was no mention of these legislative developments in AIP Italia in force at the time of the Cagliari accident. This observation led the Investigation Agency to draw up a special recommendation.[33]

From the different point of view of the Court, the old note No. 41/8880 was not only into force but was also applicable to air traffic controllers.

Actually, for many years, these Italian rules were initially unknown to the aviation personnel. They first appeared in the AIP Italia in 1996, but according to the public prosecutor's opinion, this inclusion was aimed exclusively to make these rules binding for pilots.

On the contrary, since controllers have the right and the power to authorize or refuse a visual approach, the judges decided that these rules were also addressed to them automatically.

Therefore, according to the judges' reasoning, in the present case, the air traffic controllers did not adequately inform (failing to act) the flight crew about the presence of obstacles, which were the Sette Fratelli mountains.

In addition, in the Court's opinion, the controllers provided a misleading instruction (active behaviour) to the pilots in order to maintain the minimum altitude of 2500 ft, in a flight zone quite far from the airport, where the minimum altitude was considerably higher than was indicated by the ATCs, due to the presence of the mountains. According to the judges' scenario, the controllers failed to warn the pilot (failure to act) when it became clear that the aircraft was going down at an altitude lower than the mountains.[34]

Furthermore, in the grounds of the ruling, the Supreme Court drew attention to the visual approach clearance, given by the controllers when the aircraft was still far from the airport runway (32 miles from Elmas airport). The aircraft collided with the 'Baccu Malu' mountain when it was still 17 miles from the Elmas. Therefore, the controllers, even though they had realized that the aircraft was in a dangerous, abnormal position, did not warn the pilots of danger (omission).

Finally, the controllers were accused of violating the requirements imposed by the ICAO Doc 4444.

Article 6(5.3.3) of this document adds that: "An IFR flight may be cleared to execute a visual approach provided the pilot can maintain visual reference to the terrain".

According to the Court, in the present case, this condition had not been observed by the ATCs: due to the remoteness of the aircraft from the airport, an immediate landing was technically unachievable.

In conclusion, according to Article 43 of the Italian Criminal Code, mentioned above, this ruling lays both general and specific blame on the controllers. The first one consists in the negligent, imprudent and unskillful behaviour of the defendants, in terms of infringement of professional standards and expectations. The second one

---

[33]Recommendation ANSV-3/28-04/6/A/09.

[34]Mountains tall more than 1000 m.

consists in non-compliance with the rules governing visual approach, adequate obstacle clearance and instructions on the descent of the plane.

The judges apportioned more responsibility, new tasks and professional duties to the controllers. In fact, this ruling recognised the controller's responsibility in failing to ascertain technical knowledge, experience and skills of the cabin crew. The judges considered air traffic controllers as airway police officers, responsible for aviation safety, able and obliged, at the same time, to impose discipline to ensure the safe and orderly aircraft movements, both on the ground and in the air. The latter function implies the power of the controller to give orders and clearance to ensure the aviation safety. His position of 'guardian' of flight safety is afforded to him in order to prevent collision between aircraft and air disasters.[35]

The case-law under examination is a clear example of the difficulties to reconcile international safety rules and national law on criminal liability. There is a need for a balance between safety improvement and the interest of the judicial system, in order to provide an equitable interpretation of the law and prosecution.

5.2   The Consequences of the Supreme Court Judgement on the 2004 Cagliari Fatal Accident

Following this Supreme Court judgement, there has been a growing concern on the part of aviation professionals, including Air Navigation Service Providers (ANSPs) and representative bodies of aviation personnel such as IFATCA (International Federation of Air Traffic Controller's Associations), IFALPA (International Federation of Air Line Pilot's Associations) and ANACNA (the Italian National Association of Assistants and Controllers)[36] about the courts' interpretation of the safety position of front-line operators, which led to an abnormal unjustifiable extension of the liability of controllers.

In response to the Cagliari accident case-law, the Italian navigation service providers (ENAV and Air Force) decided to refuse clearance for the visual approach to protect their air traffic control operators.

The Cagliari disaster ruling demonstrated the lack of will of Italian jurisprudence to comply with suggestions and guidelines from the most important aviation organizations, such as ICAO, and also with European directives which were aimed to promote a 'just culture' and to create a trust environment, free from the fear of

---

[35]See Uiltrasporti, Tavolo tecnico sulle competenze del controllore del traffico aereo. Concept Document 'per proposte di modifica legislativa e regolamentare' (2011), in www.uiltrasporti.it.

[36]It is the only technical-professional no-profit association of air traffic control in Italy that does not play any political role. The Association collaborates with all organizations operating in the field of safety and efficiency of air navigation. It aims to the development of the means and procedures for a safe and expeditious flow of air traffic, in the national and international field.

punishment, in which a regular and continuous exchange of safety information is possible.

5.2.1   The ANSV Recommendations No. 3/28-04/6/A/09 and No. 4/28-0476/A/09

As has been said, a safety investigation on the Cagliari accident was conducted by ANSV. It led to recommendations that were contained in the final report.

To this purpose, it may be appropriate to clarify that 'safety recommendation' means "a proposal of a safety investigation authority, based on information derived from a safety investigation or other sources such as safety studies, made with the intention of preventing accidents and incidents".[37]

A safety recommendation in no case should create a presumption of blame or liability for an accident, serious incident or incident.

Such recommendations, addressed to the aviation authorities, contain any preventive actions considered necessary to enhance aviation safety. Each entity receiving a safety recommendation, including the authorities responsible for civil aviation safety at the national and EU levels, shall implement procedures to monitor actions taken in response to these safety proposals.[38]

Among the other safety recommendations made by the Italian Investigation Agency with regard to the Cagliari disaster, the most relevant is certainly No. 3/28-04/6/A/09. This was where ANSV, after having observed that the national regulatory framework was nebulous and AIP Italia did not include legislative developments, suggested to ENAC, Air Force and ENAV that they establish a regular information and monitoring system, contained in AIP Italia, paying particular attention to regulations and procedures, in order to avoid non-compliance with existing legislation.

In particular, it was recommended by the Investigation Agency to ensure that information related to regulations and procedures, contained in AIP and in operating manuals, do not give rise to doubts of interpretation, which would lead to a negative impact on safety.

Another very important recommendation is No. 4/28-0476/A/09, in which the Agency observed that the then-in- force service order No. 10 of 8 August 2001 stated that aircraft cleared for a visual approach should have been instructed to maintain an altitude of not less than 2500 ft, pending another clearance from the Elmas aerodrome controller tower (TWR), whereas permanent internal instructions, issued by

---

[37]Article 2, point 15, of Regulation (EU) No. 996/2010, mentioned above.

[38]So Regulation 996/2010 on the investigation and prevention of accidents and incidents in civil aviation. In: https://www.skybrary.aero/index.php/Regulation_996/2010_on_the_investigation_and_prevention_of_accidents_and_incidents_in_civil_aviation. See Horspool and Humphreys (2012); Jakab and Kochenov (2017). See also Pescatore and Capotorti (1987); Cremona (2018).

Air Traffic Services Office of Cagliari airport and in force at the time of the accident, stated that Cagliari APP, before authorizing a visual approach below 3000 ft, should have received clearance from the Elmas TWR.

ANSV also recommended the Air Force and ENAV to ensure that rules on the air traffic service provision, originated by different entities, are consistent and coordinated, especially if they refer to surrounding airspaces.

5.2.2   The NOTAM A/4785

Faced with this situation, in the aftermath of the accident, ENAC enacted a provision that led to the issuing of the NOTAM[39] No. A/4785 of 25 August 2004.

This notice annulled, with immediate effect, the paragraph 9.1.3[40] of the above RAC 1-48, replacing it with the following text on 'night visual approach': "Pending the enactment of new requirements by ENAC, during the night, a visual approach shall be prohibited, in national territory, for general air traffic, as defined in ICAO Annex 6".

A note to his provision says: "A visual approach for commercial air transport and aerial work shall comply with the requirements imposed by the State of the aircraft operator".[41]

This is a significant change compared to the original text, because any reference to the notes No. 41/8879 and No. 41/8880, examined above, has been deleted and, pending the enactment of new requirements, full compliance with conditions imposed by the State of the operator has been required.

The Italian Civil Aviation Authority thus demonstrated that it wanted to give an authentic interpretation concerning the subjective scope of these two notes, finding them certainly inapplicable to the controllers.

5.2.3   The ENAC Circular ATM-07

Then ENAC held a technical board meeting, involving all stakeholders, on the possible modification of the legislation in force, with the task of producing a

---

[39] A Notice to Airmen (NOTAM) is a notice filed with an aviation authority to alert aircraft pilots of potential hazards along a flight route or at a location that could affect the safety of the flight. NOTAMs are created and transmitted by government agencies and airport operators under guidelines specified by ICAO Annex 15 (Aeronautical Information Services) of the Convention on International Civil Aviation.

[40] Para, 9.1.3: "During the night, a visual approach shall be permitted, in the national territory, only for commercial air transport and aerial work under the conditions laid down by ENAC in the provisions No. 41/8879/AM.O. and No. 41/8880 AM.O., both dated 20 June 1991".

[41] See point 1.10.

legislative text aimed at better clarifying duties and responsibilities of air traffic controllers, in accordance with the principles of 'just culture'.

Consequently, on 5 November 2013, ENAC published the Circular[42] ATM-07[43] on "Visual approach: allocating tasks and responsibilities", that specifies roles and responsibilities of the pilots and ATC personnel in case of visual approach in the Italian airspace.

Article 6(2) of this circular clarifies that the visual approach procedure is totally undertaken by the commander, from start to finish, and is conducted under his full responsibility. Therefore, the pilot in command executes the proper performance of the manoeuvres and their safe completion, constantly ensuring a safe terrain clearance and adequate separation from the obstacles that are met along the approach route. That is made quite clear in the text: "It does not fall within the ATC's area of competence to provide information about the terrain", with reference to the relief of the area flown over.

Whenever deemed necessary in the interest of air traffic safety, the ATC personnel has the power to impose temporary restrictions on the descent of aircraft using visual approach procedures. These restrictions need to be amended as quickly as possible.

Based on findings in its Safety Management System, the navigation service provider may propose that eventual local requirements or restrictions following consultation of the parties concerned be published in AIP Italia.

Article 7 of the Circular under examination says that, when requested by the pilot in command, clearance for visual approach can be issued by the air traffic controller for many reasons: for example, to prevent the negative effects of potentially hazardous weather conditions.

Visual approach can be only proposed by the controller during the day and for air traffic control reasons, subject to the consent of the pilot in command.

In both cases, whether the pilot in command is authorised by the controller or gives him his consent, he shall assume full responsibility concerning the existence of all conditions for a 'visual approach'.

It is not necessary to go into any more detail on the matter. It is appropriate to mention only those provisions of the Circular ATM-07 that demonstrate the effort of the Italian regulator to apply 'just culture' principles to the sector of air traffic control and to avoid other judgments and judicial decisions, such as that relating to the Cagliari disaster, based on 'blame culture' principles.

The rulings that sentenced the two military controllers who did their duty and complied with applicable rules and procedures are clearly inspired by a 'blaming culture', aimed to find a scapegoat, no matter what.

---

[42]ENAC circulars contain interpretations and methods of compliance with regulatory rules. The applicability of a circular to the different operators shall derive from its content.

[43]ENAC Circular ATM-07, Series ATM (Air Traffic Management) of 5 November 2013, entitled "Avvicinamento a vista: compiti, attribuzioni e responsabilità".

Simply think of the consequences that could arise as a result of public and media pressures[44] that generally accompany any crash or serious incident, with the search for culprits and easy scapegoats.

The Circular under examination, in fact, aims to draw the dividing line between the responsibility of the controller and that of the pilot in command. The latter remains fully responsible for the maneuvers of the aircraft, ensuring a safe terrain and obstacle clearance. It falls within the controller's tasks to prevent collisions between aircrafts in flight and on the ground, in the airport manoeuvring area, and obstructions in that area. Therefore, it does not fall within the ATC's area of competence to provide information about the terrain.

5.2.4   The Italian Legislative Decree No. 133/2014

The criticisms raised by the Court of Cassation's judgement on the Cagliari crash led to Article 28 of the Law Decree No. 133/2014 bringing about amendments and additions to the Italian Navigation Code.[45]

In particular, the following changes were made to the Navigation Code: (a) to Article 691-*bis*,[46] fourth paragraph, first period, relative to the provision of air traffic services by Air Force, the following phrase was added: "also in order to ensure, in the provision of air navigation services, a level equivalent to that provided for in the Community legislation". The intention to ensure a high standard of safety in the provision of air navigation services, in particular in the European skies, is very clear; (b) the new Article 733-*bis* (Tasks of pilots, pilots in command and personnel engaged in providing air navigation service for general air traffic) has been introduced.

This new Article states that

> Tasks, powers and relative operating procedures of the aircrew referred to in Article 732, first paragraph, letter a and of ground crew referred in Article 733, first paragraph, letter a and of the military personnel providing air navigation services for general air traffic shall be governed by European law, and by national technical regulations adopted by ENAC under Articles 687, first paragraph, and Article 690, first two paragraphs, as well as by the implementation manuals of air navigation services providers, Air Force and airline operators.

---

[44]Koivu (2013), p. 67.

[45]Law Decree No. 133 of 12 September 2014 laying down *"Urgent measures for the opening of the construction sites, the execution of public works, the digitization, the bureaucratic simplification, the emergence of hydrogeological instability and the resumption of production activities"* (G.U. No. 212 of 12 September 2014), converted with amendments into the Law No. 164 of 11 November 2014.

[46]As modified by Article 2 of the Legislative Decree No. 151 of 15 March 2006 and by Article 28, point 8 letter a of the Law Decree No. 133 of 12 September 2014. The first one is the Law reforming the Navigation Code, restricted to the aviation part.

The aim of this new Article is to confine the scope of legislation and binding regulation on air traffic controllers involved in general air traffic operations within the boundaries of the referred legal framework, thus avoiding the possibility that these personnel might be charged for infringements of non-mandatory requirements, as was the case in the Cagliari disaster. In particular, these legislative modifications have been introduced in order to avoid blaming ATC personnel for pilots' errors.

5.3   The Italian Legal System Following the Changeover from a 'Blaming Culture' to a 'Just Culture'

With regard to the air transport sector, the Italian legal system has been strongly affected by the European legislation, which has had a considerable impact also regarding the changeover from a 'blame culture' to a 'just culture'.

In fact, the process that began with the implementation of the already mentioned Council Directive 94/56/EC (establishing the fundamental principles governing the investigation of aviation accidents and incidents) by the Legislative Decree No. 66/1999, continued to grow in subsequent years.

This Legislative Decree, and also in relation to Directive 94/56 itself pursued the aviation safety objectives following a reactive approach by means of investigation, as it was applied after an adverse event. The subsequent Directive No. 2003/42/EC on occurrence reporting in civil aviation, implemented by the Legislative Decree No. 213/2006, both mentioned above, followed a proactive approach by utilizing preventive measures (the reporting system) aimed at eliminating the risk of accidents or incidents.

Article 3 of this Decree introduced a legal obligation on certain natural and legal persons to report occurrences listed in Annexes 1 and 2.

ENAC, through an appropriate reporting system, was responsible for collection, analysis and dissemination of this data for statistical and safety purposes. This data analysis allowed the detection of deficiencies that, once removed, led to the reduction of accidents and incidents.

The same Decree also assigned the task of establishing and managing a voluntary reporting system to ANSV.

The aforementioned legislation has been repealed and replaced by Regulation (EU) No. 376/2014, mentioned here on many occasions and directly applicable in all member States. It introduces a regulatory and harmonized framework on reporting, analysis and follow-up of occurrences in civil aviation.

We have seen that the occurrence reporting system is the principal instrument to ensure a 'just culture'.

Subsequently, this regulatory framework allowed the transposition of 'just culture' principles into Italian law, while the repealed legislation did not expressly refer to these principles, the existing Regulation No. 376/2014 expressly relates to it. Actually, Article 2(12), contains the common definition of 'just culture'.

In particular, we remember that recital 36 of this Regulation stresses that the civil aviation system should promote a 'safety culture', facilitating the spontaneous reporting of occurrences and "thereby advancing the principle of a just culture". In addition, the subsequent Article 16(6), albeit with reservations ("without prejudice to applicable national criminal law"), states that "Member States shall refrain from instituting proceedings of unpremeditated or inadvertent infringements of the law which come to their attention only because they have been reported pursuant to Articles 4 and 5".

5.4 From the (EC) Regulation No. 216/2008 to the Italian Legislative Decree No. 173/2017: A Step Backwards in the Framework of Legislation Based on 'Just Culture' Principles?

As regards the principles of 'just culture', after the de-evolution recorded in the healthcare sector as a result of the entry into force of the Italian 'Gelli' Law,[47] Law No. 173 of 15 November 2017[48] also represents a step backwards in the aviation field.

The latter Decree establishes the penalties system for infringements of the provisions contained in (EC) Regulation No. 216/2008 (so-called 'Basic Regulation').[49]

This Regulation introduced specific common rules in the field of civil aviation with the aim to achieve various objectives, the most important of which is to establish and maintain a high uniform level of civil aviation safety in Europe. In many of its recitals and Articles this objective has been pointed out very well.

It is true that this text did not yet expressly refer to 'just culture' principles, considering that, as we have seen, the first European framework based on these principles is the Regulation No. 691/2010. However, recital 16 of the European Regulation of 2008 reads as follows: 'The promotion of a 'safety culture' and the proper functioning of a regulatory system in the fields covered by this Regulation require that incidents and occurrences be spontaneously reported by the witnesses thereto'. It also clarifies 'Such reporting would be facilitated by the establishment of

---

[47]Cited above, chapter 3, para. 3.1.

[48]Legislative Decree 15 November 2017, No. 173 laying down sanctions regime in case of infringements of the provisions contained in EC Regulation No. 216/2008 on common rules in the field of civil aviation (G.U. No. 284 of 5 December 2017). For a commentary see Taviano (2018), p. 245 ff.

[49]Regulation (EC) No. 216/2008 of the European Parliament and of the Council of 20 February 2008, on common rules in the field of civil aviation and establishing a European Aviation Safety Agency, and repealing Council Directive 91/670/EEC, Regulation (EC) No. 1592/2002 and Directive 2004/36/EC (O.J. L 79 of 19 March 2008).

a non-punitive environment, and appropriate measures should be taken by Member States to provide for the protection of such information and of those who report it'.

Although this recital makes explicit reference to a 'safety culture' (and not uses the term 'just culture), it clearly refers to 'just culture' principles, especially where it requires the establishment of a 'non-punitive environment', in which incidents and occurrences can be spontaneously reported.

In addition, with reference to a non-punitive environment, Article 16(2) of the same Regulation says: "Without prejudice to applicable rules of criminal law, Member States shall refrain from instituting proceedings in respect to unpremeditated or unintentional infringements of the law which come to their attention only because they have been reported pursuant to this Regulation and its implementing rules".

This provision, subsequently repeated by Regulations (EU) Nos. 376/2004 and 996/2010 as we have seen, is closely linked to recital 16: to promote a non-punitive environment, in which aviation operators are induced to spontaneously report occurrences and other deficiencies; it implies the obligation of Member States to refrain from prosecuting aviation operators in case of 'unpremeditated or unintentional infringements' of the law.

We wonder why the Regulation under examination does not expressly refer to a 'just culture'. At the time when this Regulation was adopted, at the international level, the 36th ICAO Assembly had already taken place and the relative Resolutions, incorporating 'just culture' principles, had already been issued. We can not think that, in 2008, when the Basic Regulation was adopted, the European Union ignored the comprehensive studies already carried out by Eurocontrol on this issue.

For all these reasons, we can think that Regulation No. 216/2008 has been implicitly inspired by 'just culture' principles.

Article 68 of this Regulation expressly requires: "Member States shall lay down penalties for infringement of this Regulation and its implementing rules. The penalties shall be effective, proportionate and dissuasive".

Italian legislature implemented the latter provision by means of Legislative Decree No. 173/2017, which set out administrative sanctions applicable in case of infringement of Articles 5, 6, 7, 8, 8-*ter*, 8-*quarter*[50] of the 'Basic Regulation' No. 216/2008 and its implementing rules.

The main aim of the present analysis is to examine the new Italian law more closely, establishing penalties for infringements of Regulation No. 216/2008, in order to check its compatibility with 'just culture' European principles.

---

[50] Articles 8-*ter* and 8-*quater* have been introduced by the Regulation (CE) No. 1108/2009 of the European Parliament and of the Council of 21 October 2009 amending Regulation (EC) No. 216/2008 in the field of aerodromes, air traffic management and air navigation services and repealing Directive 2006/23/EC (O.J. L 309 of 24 November 2009, 51). For a commentary, see Vincenzi (2010), p. 316 ff.

Furthermore, it must be quite clear that our analysis will be focused on Article 8-*ter* of this Regulation, concerning the applicable rules of the air,[51] and not also include other provisions, involving certification requirements. This Article, in turn, refers to the Annexes V *bis* and V *ter* of the same Regulation.

In fact, it states that "Provisions of ATM/ANS shall comply with the essential requirements set out in Annex Vb and, as far as practicable, Annex Va". Annex Va (or *bis*) refers to "Essential requirements for aerodromes", while Annex Vb (or *ter*) refers to "Essential requirements for ATM/ANS and Air Traffic Controllers".

Some clarification is needed on this matter.

There are no provisions that define penalties applicable for violations of EU law. The nature and level of such penalties are determined by the national law of the Member States.

Therefore, the nature of national penalties varies widely from one EU Member State to another. For the same violations, some of them choose a penalty regime while others rely on administrative penalties. Some Member States adopt a combination of the two systems.

When Member States set and apply penalties for infringements on EU law provisions, they are free to choose the measures considered most appropriate in terms of nature and severity, but—and it is important to remember this—in accordance with the principles of EU law.[52]

In the famous *Commission/Greece* case law,[53] the European Court of Justice (CJEU) clearly identified an obligation of Member States, contained in Article 10 of the former EC Treaty,[54] now Article 4(3)[55] of the EU Treaty (TEU), to penalize those who breach EU law. The Court stressed that, in accordance with the principle (*rectius*, mutual duties) of sincere co-operation[56] contained in these Articles, Member States are obliged to ensure that EU law is effectively applied and enforced at national level. In particular, with respect to infringements of Community law, the national authorities are required to use the same level of diligence required in implementing corresponding national laws.

---

[51]The rules of the air are subject to the ICAO Annex 2, entitled "Rules of the Air", edition July 2005. See Milde (2008), p. 69.

[52]World Trade Organization, Dispute Settlement Reports, 2006, Vol. IX, Cambridge University Press, Cambridge, 2008, p. 3958.

[53]Judgement of the Court of Justice of 21 September 1989, Commission of the European Communities v Hellenic Republic, case C-68/88, in ECR (European Court Reports), 1989, p. 2965, para. 22-25. For a commentary, see Honorati (2006), p. 941 ff.; Miettinen (2013), p. 13. Similarly, Hansen, case C-326/88, in ECR (European Court Reports), 1990, I-2911, para. 17; Siesse, case C-36-94, in ECR (European Court Reports), 1995, I-3573, para. 20; Andrade, C-213/99, in ECR (European Court Reports), 2000, I-11083, para. 19–20.

[54]Temple Lang (1986), p. 503; Costantinesco (1987), p. 97 ff.; Blanquet (1994); Nizzo (1997), p. 381 ff.; Mortelmans (1998), p. 67 ff.

[55]More recently, see Larick (2018), part II, para. 7.

[56]See also Bakker et al. (1995), p. 142.

This judgement clarifies that, even though the choice of penalties remains within the discretion of the Member States, they must ensure: (a) that infringements of EU law are punished under (substantive and procedural) conditions similar to those applicable to infringements of national law of the same nature and relevance,[57] and (b) that penalties provided for under national legislation are effective, proportionate and dissuasive.[58]

Many European regulations (such as the 'Basic Regulation' No. 216/2008 under examination) or directives request that Member States impose "effective, proportionate and dissuasive penalties" on violations of their provisions, but they do not specify the way in which they must do so and what this notion precisely means.

Before carrying out a comparison between Article 68 of the European 'Basic Regulation' and the new Italian Legislative Decree of 2017, we need to ask the following basic question: what is the meaning of the phrase 'effective, proportionate and dissuasive penalties' from a general EU law point of view?

Actually, this notion derives from the case-law of the European Court of Justice.

There is no doubt that penalties are 'dissuasive' where they serve as a deterrent to prevent an individual from infringing on the objectives of EU law'.[59] In other words, the sanctions must constitute an adequate deterrent for potential future perpetrators.

According to the CJEU's point of view, in order to make penalties 'effective' "Member States must ensure that infringements of EU law are penalised under conditions analogous to those applicable to infringements of national law of similar nature and importance".[60]

This notion has been so interpreted by the Court: "an effective penalty is understood as one that ensures that the same diligence and strictness is applied for breaches of EU law as for breaches of provisions stemming from national law".[61]

In addition, under another opinion, the effectiveness[62] requires that penalties should not only achieve deterrence, but also aims at restoring the damage caused and preventing future risk.[63]

---

[57]CJEU, judgement of 7 October 2010, *Stils Met*, C-382/09, EU:C:2010:596, par 44.

[58]See Udvarhelyi (2016).

[59]So the Opinion of the Advocate General Kakott in the joined cases C-387/02, C-391/02 and C-403/02, *Berlusconi and others*, in ECR (European Court Reports), 2005, I-3565. For a commentary, see Chalmers et al. (2010), p. 215; Cagnazzo et al. (2012).

[60]See CJEU judgement of 7 October 2010, *Stils Met*, C-382/09, EU:C:2010:596, par. 44.

[61]See National EUTR penalties: are they sufficiently effective, proportionate and dissuasive?, in ClientEarth, March 2018. In: www.documents.clientearth.org.

[62]In the Opinion of the Advocate General Kakott, mentioned above, penalties are considered 'effective' where they are such that they do not make it practically impossible or excessively difficult to impose to achieve the objectives pursued by the EU law.

[63]In particular, with reference to the environment law. See Vagliasindi (2017), chapter 3; Faure (2010), p. 256 ff.; Fitzmaurice et al. (2016), p. 33 ff.; Farmer et al. (2017).

In EU soft law instruments,[64] instead, the European Commission clarified that effectiveness requires that the sanction is suitable to achieve the desired goal, *i.e.* observance of the rules and principles.

Nevertheless, what is the real meaning of the term 'proportionate'? In our opinion, it has at least two different meanings.

In general terms, within the EU law, penalties are 'proportionate' where they are appropriate for achieving objectives and complying with principles of the EU law.[65] In other words, a close relationship exists between means and an end (proportionality).

In addition, according to the general principle of proportionality in criminal law, the punishment should fit the crime.

In penal law, the principle of proportional justice is used to describe the idea that the punishment of a crime should be in proportion to the gravity of the crime itself.

In this respect, proportionality is interpreted by reference to a graduated punishment system. Therefore, the penalty should be proportional to the way in which particular protected interests have been endangered or infringed upon.[66]

In the EU Commission Communication entitled 'Towards an EU criminal policy: ensuring the effective implementation of EU policies through criminal law",[67] we can read: "proportionality requires that the sanction must be commensurate with the gravity of the conduct and its effects and must not exceed what is necessary to achieve the aim'.

This notion tries to reconcile the two meanings described above.

This principle, which is typical in the area of criminal prosecution, is applicable to all proceedings and sanctions, including administrative measures.

Having said that, in conformity with the European principles of effectiveness (*i.e.* compliance with the EU rules and principles) and proportionality (between the severity of the crime and the punishments), the Italian criminal law should punish only actions or omissions made intentionally ('wilful misconduct') or as a result of serious recklessness ('gross negligence').

In doing so, 'just culture' principles would be fulfilled by the Italian legislature. But that has not been the case.

Article 1 of the Legislative Decree No. 173/2017 sets out only administrative sanctions applied by ENAC (the authority responsible for applying the 'Basic Regulation' No. 216/2008) "without prejudice of the application of criminal rules for infringements of the provisions contained in Regulation No. 216/2008".

---

[64]See COM/2011/0573 final, Communication from the Commission to the European Parliament, the Council, the European Economic and Social Committee and the Committee of the Regions, Brussels, 20 September 2011, para. 3.

[65]See Schültze and Tridimas (2018), note 33.

[66]So Faure (2010), p. 256 ff.

[67]COM/2011/0573 final, Communication from the Commission to the European Parliament, the Council, the European Economic and Social Committee and the Committee of the Regions, Brussels, 20 September 2011, cited above, para. 3.

The last parenthesis in the text implies that the 'blame culture' approach, based on Article 43 of the Italian Penal Code, which does not recognize levels of fault, shall continue to be followed in criminal proceedings.

But also in respect to administrative penalties, the Decree under examination does not follow the 'just culture' approach. The administrative fines laid down in it are set out without taking into account the distinction between honest violations and deliberate or seriously negligent infringements.

Actually, Article 4(2) of this Decree specifies that it will comply with the general provisions of chapter I, sections I and II, of Italian Law No. 689/81,[68] *mutatis mutandis*.

Article 11 of the Italian Law No. 689/81 (the so-called "decriminalisation Law"), that regulates proceedings other than criminal ones, reads as follows: "in determining the amount of the administrative fines, set between a minimum and a maximum, the infringement severity shall be taken into account".

Notwithstanding the statutory reference to the criterion of the severity of the violation, Law No. 689/81 does not distinguish between 'gross negligence' and 'honest behaviour', but leaves room for broad discretion to the competent authorities in ascertaining the effective gravity of the infringement in order to impose sanctions.

The Italian Legislature should impose administrative sanctions only in the case of 'gross negligence' or 'wilful misconduct' rather than refer, on this point, to the Law of 1981, which is not in line with 'just culture' principles.

Instead, the recent Decree No. 173/2017 contains no clear provision excluding punishability for inadvertent violations and genuine mistakes.

In fact, in the case of minor breaches, Article 4(3) is limited to provide for a formal injunction by ENAC, addressed to the offender, in order to remedy such slight violations through required corrective measures, within a specified period.

Some authors[69] rightly observed that this Article, as well as Article 11 of the Law No. 689/81, refers to the objective element of the violation, *i.e.* to the fact that it has been committed and not to the subjective element (degree or level) of the blame, as a 'just culture' approach would require.

Out of these cases, Article 12 punishes front-line operators who violate the rules of the air, even in the case of honest mistakes.

Airline pilots, air traffic service personnel, meteorological service providers and ground personnel engaged in aircraft operations are required to pay a fine ranging from 1000 to 10,000 euros in case of violation of the rules of the air contained in Article 8-*ter* of the 'Basic Regulation' and referred to 'applicability and compliance, general rules and collision avoidance, signals, time, flight plans, visual meteorological conditions, visual flight rules and instrument flight rules, airspace classification, air navigation service, air traffic control service, flight information service, alerting

---

[68]Law No. 689 of 24 November 1981 "Changes to the penal system" (G.U. No. 329 of 30 November 1981). For a commentary, see Stravino (2012), p. 877 ff.

[69]Taviano (2018), p. 261.

service, interference, emergency contingencies and interception, services related to meteorology'.

This detailed list replicates what is contained in the Annex to the implementing Regulation (EU) No. 923/2012,[70] entitled "Rules of the Air" (SERA).

Article 8-*ter*, however, is formulated in a very generic way, not specifying the connection between violations and monetary penalties.[71]

No provision of the Decree takes into account the distinction between acceptable and unacceptable behaviour of the front-line operators.

This text should have taken into account the unpremeditated or unintentional infringements, rather than to refer to their gravity.

No provision of the Decree emphasises the difference between violations that have been reported and those obtained from other sources, considering that no protection is given to the reporters, in contrast to Article 16 (as well as recital 16) of the 'Basic Regulation'.

Therefore, under examination, Article 12 of the Decree is also in contradiction with Articles 16(2) (as well as recital 16) of the same Regulation that should have fully implemented the 'safety culture' (*rectius*, 'just culture') principles because it did not respect the spirit of the European text.

Member States have a duty to ensure that EU law is effectively applied and enforced at national level in accordance with the principle of sincere cooperation laid down in Article 4(3) of the TUE as mentioned above.

Finally, the Italian Decree is in contrast with Article 68 of the same 'Basic Regulation', considering that it did not introduce proportionate penalties, *i.e.* penalties commensurate with the gravity of the conduct.

Therefore, the Italian State has not complied with the principle of loyal cooperation (or fidelity principle)[72] and for this reason it may be subject to an infringement procedure provided for in Article 258 of the Treaty on the Functioning of the European Union (TFEU).[73]

However, over and above that, the Decree No. 173/2017 represents a step backwards in the framework of (international, European and national) legislation based on the application of 'just culture' principles in the aviation sector.

---

[70]Commission Implementing Regulation (EU) No. 923/2012 of 26 September 2012 laying down the common rules of the air and operational provisions regarding services and procedures in air navigation and amending Implementing Regulation (EU) No. 1035/2011 and Regulations (EC) No. 1265/2007, No. 1794/2006, No. 730/2006, No. 1033/2006 and No. 255/2010 (O.J. L 281 of 13 October 2012).

[71]See ANACNA, Italian Air Traffic Controller Association, *Decreto legislativo 173/2017: più sanzioni, meno 'just culture'*, Rome, 16 December 2017. In: Comunicato_ANACNA_sanzioni_173_2017.pdf.

[72]Ippolito (2007), p. 131; Draetta (2009), p. 75; Vitale (2018), p. 15, Rossi (2018), p. 865.

[73]Chalmers et al. (2010), p. 224; Andersen (2012), chapter 2; Foster (2013), p. 160 ff., and p. 197 ff.; Gormley (2017), p. 65 ff.; Berry et al. (2017), p. 298 f.

This surprises us even more since ENAC, in its recent National Civil Aviation Safety Programme, entitled 'State Safety Programme – Italy',[74] undertook to make its safety policy compliant[75] with 'just culture' principles.

References

Alicke MD (2000) Culpable control and the psychology of blame. Psychol Bull 126(4):556–574

Andersen S (2012) The enforcement of EU law: the role of the European Commission. Oxford University Press, Oxford

Arrigoni N (1992) Joint Aviation Authorities: development of an international standard for safety regulation. The first steps are being taken by the JAA. Law and Policy in International Business

Bakker R, Heringa AW, Stroink FAM (1995) Judicial control: comparative essays on judicial review. Maklu Uitgevers, Antwerpen, p 142 ff

Barra B (2013a) La giurisprudenza dell'ultimo decennio: prevalenza della norma penale sulle regole tecniche internazionali e comunitarie; impatto su compiti e responsabilità degli operatori aeronautici. In: Bruni F, Tullio L (eds) Sinistri aeronautici: rischi e responsabilità. Giappichelli, Torino, pp 8–27

Barra B (2013b) Sinistri aeronautici: rischi e responsabilità, Seminar held within the Bar Council of Rome, 5 April 2013, in www.stasa.it

Battiati A (2011) La posizione di garanzia dei controllori del traffico aereo: responsabilità per fatto proprio o per fatto altrui? Giustizia penale 12:645–660

Berry E, Homewood MJ, Bogusz B (2017) Complete EU law: text, cases, and materials, 3rd edn. Oxford University Press, Oxford

Blanquet M (1994) L'article 5 du Traite' C.E.E. Recherce sur les obligations de fidelité des Etats membres de la Communauté. Librairie générale de droit et de jurisprudence (LGDJ), Paris

Bruni F, Tullio L (eds) (2013) Sinistri aeronautici: rischi e responsabilità. Giappichelli, Torino

Bufo M (2012) Le competenze degli ANSPs. Il servizio informazioni volo (FIS) e il servizio informazioni volo aeroportuale (AFIS). In: Pellegrino F (ed) Air navigation Rules and Practices in Europe: towards harmonization. Giuffrè, Milano, pp 163–190

Cagnazzo A, Toschei S, Pozzi C (eds) (2012) Le sanzioni in materia di trasporto marittimo, aereo, terrestre e Codice della strada. Giappichelli, Torino

Cecchi R (1993) Joint Aviation Authorities: situazione attuale e sviluppi futuri. In: Franchi B (ed) Cinquant'anni di codice della navigazione: profili di responsabilità degli operatori del settore aereo. Aero Club d'Italia, Roma, p 129 ff

Chalmers D, Davies G, Monti G (2010) European Union Law. Cambridge University Press, Cambridge, p 215 ff

Comenale Pinto MM (1997) Considerazioni sull'organizzazione dell'assistenza al volo. In: Studi in onore di Gustavo Romanelli. Giuffré, Milano, p 355 ff

Comenale Pinto MM (1999) L'assistenza al volo. Evoluzione, problemi attuali, prospettive. Cedam, Padova

Comenale Pinto MM (2007) Le competenze di ENAV S.p.A. ed i servizi della navigazione aerea. In: Franchi B, Vernizzi S (eds) Il diritto aeronautico fra ricodificazione e disciplina comunitaria. Giuffré, Milano, pp 107–110

Costantinesco V (1987) L'art. 5 CEE. De la bonne foi à la loyauté communautaire. In: Pescatore P, Capotorti F (eds) Du droit international au droit de l'integration. Liber Amicorum Pierre Pescatore. Nomos Verlagsgesellschaft, Baden Baden, p 97 ff

---

[74]ENAC, State Safety Programme – Italy, edition 3, February 2017.

[75]See the definition of 'safety policy', 8 and lett. j), p. 9.

Cremona M (ed) (2018) Structural principles in EU external relations law. Bloomsbury, London

Draetta U (2009) Elementi di diritto dell'Unione Europea. Parte istituzionale. Ordinamento e struttura dell'Unione Europea. Giuffré, Milano

Farmer A, Faure M, Vagliasindi GM (eds) (2017) Environmental crime in Europe. Bloomsbury, London

Faure M (2010) Effective, proportional and dissuasive penalties in the implementation of the environmental crime and shipsource pollution directives: questions and challenges. Eur Energy Environ Law Rev 19(6):256–278

Federal Aviation Administration (2017) Instrument procedures handbook. Skyhorse Publishing, New York

Fitzmaurice M, Martinez Gutierrez NA, Hamza R (eds) (2016) The IMLI manual on international maritime law, vol. III, Marine Environmental Law and Maritime. Oxford University Press, Oxford

Foster N (2013) Foster on EU law. Oxford University Press, Oxford

Franchi B (ed) (1993) Cinquant'anni di codice della navigazione: profili di responsabilità degli operatori del settore aereo. Aero Club d'Italia, Roma

Franchi B, Vernizzi S (eds) (2007) Il diritto aeronautico fra ricodificazione e disciplina comunitaria. Giuffré, Milano

Freni E (1997) La trasformazione dell'azienda di assistenza al volo in ente pubblico economico. Giornale di diritto amministrativo 7:605 ff

Gallo G (2002) A proposito di visual approach. In: http://www.traffico-aereo.it/pages/atcp/atcp07/atcp07.htm

Gibb R, Gray R, Scharff L (2016) Aviation visual perception, research and mishaps. Ashgate studies in human factors for flight operations. Taylor & Francis Group/Routdlege, London/New York

Gormley LW (2017) Infringement proceedings. In: Jakab A, Kochenov D (eds) The enforcement of EU law and values: ensuring member states' compliance. Oxford University Press, Oxford. part II, para. 4, pp 65 ff

Grigoli M (1997) Profili della evoluzione normativa e sistematica dell'assistenza al volo in Italia. Trasporti (72-73):23–42

Honorati C (2006) La comunitarizzazione della tutela penale e il principio di legalità nell'ordinamento comunitario. Rivista di diritto internazionale privato e processuale:941–986

Horspool M, Humphreys M (2012) European Union Law. Seventh edition core text series. Oxford University Press, Oxford

Hunter DR, Baker RM (2000) Reducing accidents among general aviation pilots through a national Aviation Safety Program. In: The Fourth Australian Aviation Psychology Symposium, Aldershot, Ashgate (England)

Iovino A (2013) Air cops and mountain tops. HindSight 18:78–81

Ippolito F (2007) Fondamento, attuazione e controllo del principio di sussidiarietà nel diritto della Comunità e dell'Unione Europea. Giuffré, Milano

Jakab A, Kochenov D (eds) (2017) The enforcement of EU law and values: ensuring Member States' compliance. Oxford University Press, Oxford

Jing H-S, Battea A (2016) The Dragon in the Cockpit: How Western Aviation concepts conflict with Chinese value systems. Routledge/Taylor & Francis, London/New York

Koivu H (2013) Just and safety: the art of making mistakes. People sell washing machines – robots fly aeroplanes? Hindsight 18:66–69

La Penna V (1992) L'Azienda nazionale di assistenza al volo: ente pubblico o azienda autonoma? Rivista della Corte dei conti 4:182–195

La Penna V (1993) L'Azienda nazionale di assistenza al volo: ente pubblico o azienda autonoma? in Enti pubblici, pp 379 ff

La Torre U (2015a) Aircraft pilotage. Ordines 1:2–28

La Torre U (2015b) Aircraft pilotage. In: Pellegrino F (ed) Legislation and regulation of risk management in aviation activity, vol II. Giuffré, Milano, pp 3–25

La Torre U (2016) Il "visual approach" notturno nella giurisprudenza italiana. Rivista diritto navigazione 1:75–105

Larick J (2018) Pars pro toto: the Member States' obligations of sincere cooperation, solidarity and unity. In: Cremona M (ed) Structural principles in EU external relations law. Bloomsbury, London

Miettinen S (2013) Criminal law and policy in the European Union. Taylor & Francis, Abingdon

Milde M (2008) International Air Law and ICAO. Eleven International Publishing, Utrecht

Mortelmans K (1998) The principle of loyalty to the community (Article 5 EC) and the obligations of the community institutions. Maastricht J Eur Comp Law 5:67–88

Nizzo C (1997) L'art. 5 del Trattato CE e la clausola di buona fede nell'integrazione europea. Diritto dell'Unione Europea:381–392

Oxford Aviation Academy (2008) Air law, ATPL ground training series. Oxford Aviation Academy, Oxford, p 145

Pellegrino F (2007) Sicurezza e prevenzione degli incidenti aeronautici. Giuffrè, Milano

Pellegrino F (ed) (2012a) Air navigation rules and practices in Europe: towards harmonization. Giuffrè, Milano

Pellegrino F (2012b) Il contratto di trasporto aereo e la sanzione amministrativa. In: Cagnazzo A, Toschei S, Pozzi C (eds) Le sanzioni in materia di trasporto marittimo, aereo, terrestre e Codice della strada, vol II. Giappichelli, Torino, pp 328–352

Pellegrino F (2013) Just culture principles. Aviation Law from a European perspective. Ann Air Space Law:471–490

Pellegrino F (ed) (2014) Legislation and regulation of risk management in aviation activity, vol I. Giuffré, Milano

Pellegrino F (ed) (2015) Legislation and regulation of risk management in aviation activity, vol II. Giuffré, Milano

Pescatore P, Capotorti F (eds) (1987) Du droit international au droit de l'integration. Liber Amicorum Pierre Pescatore. Nomos Verlagsgesellschaft, Baden Baden

Pooley E (2013) The accident investigation. HindSight 18:71–73

Pugliese SAF (1989) Competenze civili e militari nell'esercizio dei servizi di assistenza al volo. Diritto e pratica dell'aviazione civile:232 ff

Romanelli G (1994a) Assistenza al volo: da attività di polizia della navigazione a prestazione di servizio. In: Romanelli G (ed) Spunti di studio sull'attività di assistenza al volo, Cagliari, pp 5 ff

Romanelli G (ed) (1994b) Spunti di studio sull'attività di assistenza al volo, Cagliari

Rossi LS (2018) 2, 4, 6 (TUE)... l'interpretazione dell'"Identity Clause" alla luce dei valori fondamentali dell'UE. In: Codinanzi M, Cannizzaro V, Adam R (eds) Liber Amicorum in onore di Antonio Tizzano: De la Cour CECA à la Cour de l'Union: le long percours de la justice européenne. Giappichelli, Torino, pp 859–870

Scala M (2013) If it had happened in your country, what would the judgement have been? HindSight 18:80–81

Schültze R, Tridimas T (eds) (2018) Oxford principles of Europeam Union law. Oxford University Press, Oxford

Serrao F (1989) L'avvio dell'attività dell'Azienda autonoma di assistenza al volo per il traffico aereo generale. Enti pubblici, 863 ff

Sirena P (2013) Il settore aeronautico. L'elaborazione giurisprudenziale italiana. Relazione all'Incontro di Studio organizzato dalla Scuola Superiore della Magistratura, Centro Alti Studi Difesa: "La responsabilità colposa, posizioni di garanzia e profili concausali in attività complesse di organizzazioni a rischio consentito. Due esperienze a confronto, il settore aeronautico e quello sanitario", Roma, 20 novembre 2013

Sirena PA (2015a) The case law of the Italian Supreme Court of Cassation in the field of aviation accidents. In: Pellegrino F (ed) Legislation and regulation of risk management in aviation activity, vol II. Giuffré, Milano, pp 252–260

Sirena P (2015b) The Italian Supreme Court of Cassation. A case study of Aviatin Accident. The Controller The Controller Magazine, IFATCA, pp 6 ff

Smith D (2001) Controlling Pilot Error, Controlled Flight Into Terrain (CFIT/CFTT). McGraw-Hill Education, Milano

Starrantino C, Finocchiaro M (2013) The judicial aftermath. HindSight 18:74–77

Stravino P (2012) Quali sono i rapporti tra il delitto di cui all'art. 335 c.p. e l'illecito amministrativo di cui al combinato disposto dell'art. 5 l. 689/1981 e 213 comma 4 codice stradale? Foro napoletano 2-3:877–879

Taviano M (2018) Il quadro giuridico della just culture nel settore aeronautico italiano e i passi indietro nel d.lgs. n. 173/2017. Rivista di diritto dell'economia, dei trasporti e dell'ambiente, Giureta XVI:245–267

Temple Lang J (1986) Article 5 of the EEC Treaty: the emergence of constitutional principles in the case law of the court of justice. Fordham Int Law J 10(3):503–535

Udvarhelyi B (2016) Supranationality versus National Sovereignty – Linking points between EU law and national criminal law. In: MultiScience – XXX microCAD International Multidisciplinary Scientific Conference University of Miskolc, Hungary, 21–22 April 2016

Vagliasindi GM (2017) The EU environmental crime directive. In: Farmer A, Faure M, Vagliasindi GM (eds) Environmental crime in Europe. Bloomsbury, London, part I, chapter 3

Vincenzi L (2010) L'Unione europea rafforza il cielo unico. Diritto marittimo 1-2:316–319

Vitale G (2018) Diritto processuale nazionale e diritto dell'Unione europea. L'autonoamia procedurale degli Stati membri in settori a dierso livello di "europeizzazione", ed.it, Catania

Chapter 6
Final Remarks

In the light of the findings of this study, we can now stress the key objectives of a so-called 'just culture' and the obstacles that still hinder its proper implementation.

With reference to the first aspect, it is very clear that the principal goal of this culture is to prevent recurrence of aviation accidents and incidents by encouraging safety occurrence reporting and transparent investigation, as provided for under Regulations (EU) No. 996/2010 and No. 376/2014. It is a safety-orientated culture.

But this is only one side of the coin.

Another purpose of a 'just culture' is to strike a proper balance between conflicting interests: the need to improve aviation safety and the need to administer justice properly.[1]

As we have seen, the tool to promote the right balance between these conflicting interests has been identified by the above mentioned legislation in mutual understanding and advance agreements concluded between both safety and judicial authorities involved. These agreements are very important in order to train the judicial staff on technical aspects and the safety investigators on legal aspects, at the same time.

Having established these points, there is now a need to throw real light on the difficulties in implementation of 'just culture' principles into national legal systems, albeit introduced by (directly applicable) EU regulations.

In fact, most of the legal systems are still influenced by a widespread 'blame culture', that leads to the criminalisation of the operators.

The first difficulty of implementation[2] of this new culture is the lack of a clear definition.

In fact, the legal definition of 'just culture,' contained in Article 2 of Regulation No. 376/2014, which is fully in line with that drawn up by Eurocontrol and accepted

---

[1]Dekker (2012), para. 1

[2]Youngberg (2013), p. 176; Sinay (2014), p. 159.

© Springer Nature Switzerland AG 2019
F. Pellegrino, *The Just Culture Principles in Aviation Law*, Legal Studies in
International, European and Comparative Criminal Law 3,
https://doi.org/10.1007/978-3-030-23178-1_6

by ICAO, is based on the dividing line between intentional/premeditated and unintentional/unpremeditated behaviour, *i.e.* between willful misconduct/gross negligence and honest error. Despite this explanation, it does not clarify the concept of 'gross negligence'. Moreover, we do not even find it in other provisions.

Actually, even though the concept of 'gross negligence' is referred to in a number of EU instruments, we can argue that there is not yet a commonly agreed legal definition of 'gross negligence' in Europe and worldwide.

According to the existing definition of 'just culture', it is clear that nothing should prevent criminal prosecution in the event of wrongdoing or 'gross negligence' and, on the contrary, individuals (*i.e.* front-line operators) cannot be prosecuted for actions, omissions or decisions that appear to be due to an unpremeditated and inadvertent infringement of the law. In this context, the dividing line between lawful and unlawful behaviour would appear clearly defined. But it is no so simple. In fact, where exactly is the dividing line between acceptable and unacceptable behaviour in the absence of a clear definition of 'gross negligence'? It depends on what you mean by 'gross negligence'.

It is true that, despite this shortcoming, it can be assumed that this concept refers to a degree of serious disregard to an obvious risk and failure to take reasonable care as required by the circumstances. But it is also true that a clearly agreed-upon definition, contained in a binding instrument, it would be desiderable to avoid doubts over interpretation. This need is felt ever more because of the difference between common and civil law systems in this respect. The common law systems, unlike those of civil law, have always drawn a sharp distinction between negligence, however gross, on the one hand and fraud, bad faith and 'wilful misconduct' on the other.

A second question needs to be asked: who can draw this line?

The answer to the last question has to be the following: prosecutors and judges draw the dividing line between punishable and unpunishable behaviour.[3] But, in terms of 'just culture' principles, they should avail themselves of the necessary support from aviation experts, who know where the line is set between acceptable and unacceptable risk in this sector.

Another difficulty in the implementation of 'just culture' principles depends on the fact that EU legislation does not allow the usage of the final report of the safety investigation in subsequent criminal proceedings, unless the administration of justice decides otherwise. This fact demonstrates that, despite the legislative developments, judicial inquiries still seem to prevail over safety investigation.

Moreover, according to international and EU rules that follow a 'just culture' approach, incident or accident reports filed under an occurrence reporting scheme, as well as reports issued by investigating bodies or by the operators, should not be used as evidence in criminal proceedings against individuals.

Occurrence reports, technical investigations and witness declarations are all made for no other reason than to improve public transport safety. Therefore, deviating

---

[3]Wade et al. (2008), p. 101.

from this purpose and using such information in the interest of the administration of justice will harm future safety improvements.

Finally, as we have seen, aviation is a very complex system and, in the framework of a 'just culture', balanced accountability for both individuals and the organisation responsible for designing and improving systems in the workplace should be ensured.

Despite the fact that very often accidents occur because of organisational deficiencies, as appears from many trials, criminal responsibility of a legal independent entity is not yet commonly recognised at European level. In fact, criminal liability of corporate fault is not recognized in the majority of European States.

However, the sovereign administration of justice function remains the prerogative of national governments and prosecutors and must be respected.

In this context, the best way to break down the barriers to implementing a 'just culture' into national legal systems and to reconcile the conflicting interests is to adopt and implement national aviation prosecution policies that are inspired by a common European model policy.

References

Dekker S (2012) Just culture. Balancing safety and accountability. Ashgate (Hampshire), Aldershot
Sinay J (2014) Safety management in a competitive business environment. CRC Press/Taylor & Francis Group, Boca Raton/Florida
Wade M et al (2008) When the line is crossed... Paths to control and sanction behaviour necessitating a state reaction. Eur J Crim Policy Res 14:101–122
Youngberg BJ (ed) (2013) Patient safety handbook. Jones & Bartlett Learning, Burlington

References

Agich GJ (ed) (2012) Responsibility in health care. Kluwer, Dordrecht

Alagna R (2017) La controriforma della colpa penale dell'attività medica. Responsabilità civile e previdenza, pp 1466–1489

Alicke MD (2000) Culpable control and the psychology of blame. Psychol Bull 126(4):556–574

Allen MJ (2007) Textbook on criminal law, 9th edn. Oxford University Press, Oxford

Almond P (2012) Corporate manslaughter and regulatory reform. Palgrave Macmillan, London

Almond P, Colover S (2010) Mediating punitiveness; understanding public attitudes towards work-related fatality cases. Eur J Criminal 7(5):323–338

Amalberti R (2001) The paradoxes of almost totally safe transportation systems. Safety Sci 37:109 ff

Amato G (2018a) Conclusione giusta in linea con la norma e contro le negligenze. Guida al diritto 12:28–32

Amato G (2018b) Conclusione giusta in linea con la norma e contro le negligenze. Archivio giuridico circolazione assicurazione e resonsabilità 4:299 ff

Amato G (2018c) Sussiste la colpa quando è ravvisato un errore inescusabile. Guida al diritto 1:74–77

Andersen S (2012) The enforcement of EU law: the role of the European Commission. Oxford University Press, Oxford

Anderson M (2015) Behavioural safety and major accident hazards: magic bullet or short in the dark. In: Conference Proceedings, Hazards XVIII Symposium, 24 November 2004, IChemE/UMIST, Manchester

Angioni F (2006) Note sull'imputazione dell'evento colposo con particolare riferimento all'attività medica. Studi in onore di Giorgio Marinucci, II, Giuffré, Milano, pp 1279–1334

Antonini A (2015) La responsabilità nell'esercizio del volo. In: Pellegrino F (ed) Legislation and regulation of risk management in aviation activity, vol II. Giuffré, Milano, pp 45–54

Antonini A, Franchi B (eds) (2005) Diritto aeronautico a cent'anni dal primo volo. Giuffré, Milano

Arghami S, Nouri Parkestani H, Alimohammadi I (2014) Reliability and validity of a safety climate questionnaire. J Res Health Sci 14(2):140–145

Arnesen SA (ed) (1995) Interact '95. Chapman and Hall, London

Arrigoni N (1992) *Joint Aviation Authorities*: development of an international standard for safety regulation. The first steps are being taken by the JAA. Law and Policy in International Business

Badar ME (2013) The concept of Mens Rea in international criminal law: the case for a unified approach. J Int Crim Just 11(4):933–936

Baker DJ (2016) The right not to be criminalized: demarcating criminal law's authority. Routledge/ Taylor & Francis, London/New York

Bakker R, Heringa AW, Stroink FAM (1995) Judicial control: comparative essays on judicial review. Maklu Uitgevers, Antwerpen, p 142 ff

Balcerzak T (2017) A just culture? Conflict of interest in the investigation of aviation accidents. Sci J Silesian Univ Technol Series Transp 94:5–17

Ballinari L (1992) Per dolo eventuale: satira su un giudizio penale e una reclusione perpetua. Svizzera, Lugano

Barbarisi A (2017) L'onere della prova nella responsabilità sanitaria. Contratti, Contratti 2:217–230

Barnsteiner J, Disch J (2012a) A just culture for nurses and nursing students. In: Barnsteiner J, Disch J (eds) Second generation QSEN, an issue of nursing clinics of North America. Elsevier, Philadelphia, pp 407 ff

Barnsteiner J, Disch J (eds) (2012b) Second generation QSEN, an issue of nursing clinics of North America, Elsevier, Philadelphia

Barra B (2013a) La giurisprudenza dell'ultimo decennio: prevalenza della norma penale sulle regole tecniche internazionali e comunitarie; impatto su compiti e responsabilità degli operatori aeronautici. In: Bruni F, Tullio L (eds) Sinistri aeronautici: rischi e responsabilità. Giappichelli, Torino, pp 8–27

Barra B (2013b) Sinistri aeronautici: rischi e responsabilità, Seminar held within the Bar Council of Rome, 5 April 2013, in www.stasa.it

Bartoli R (2018) Riforma Gelli-Bianco e Sezioni Unite non placano il tormento: una proposta per limitare la colpa medica. Diritto penale contemporaneo 5:233–248

Bate P (1992) The impact of organisational culture on approaches to organisational problem-solving. In: Salaman G (ed) Human resource strategies. Sage, London, pp 229 ff

Battiati A (2011) La posizione di garanzia dei controllori del traffico aereo: responsabilità per fatto proprio o per fatto altrui? Giustizia penale 12:645–660

Baumgartner M (2011) The role for the human being in single European Sky. In: Calleja CD, Mendes de Leon P (eds) Achieving the Single European Sky: goals and challenges. Kluwer, Alphen aan den Rijn, pp 299–314

Baumgartner M, Schorer R (eds) (2017) Just Culture Manual for ATCO, ANSE & ATSEP. Behavior after an incident and further proceedings, SwissATCA

Baumgartner M, Chalck M, Pidgen J (2011) Just culture. In: Calleja CD, Mendes de Leon P (eds) Achieving the Single European Sky: goals and challenges. Kluwer, Alphen aan den Rijn, pp 281–298

Beauman C, Wilson C (2018) My revision notes: OCR a level law. Hodder Education, Hachette UK Company, London

Beever A (2016) A theory of tort liability. Bloomsbury, Oxford

Benkö M (ed) (2012) Essential air and space law, vol 10. Elven Law, Utrecht

Bennet S (2004) The 1st July 2002 mid-air collision over Überlingen, Germany: a holistic analysis. Risk Manag 6(1):31–49

Berlinger N (2005) *After harm*: medical error and the ethics of forgiveness. Johns Hopkins University, Batilmore

Bernabei V, Barbafina F (2015) Il Safety Management System nel campo dell'aviazione. In: Pellegrino F (ed) Legislation and regulation of risk management in aviation activity, vol II. Giuffré, Milano, pp 163–170

Berry E, Homewood MJ, Bogusz B (2017) Complete EU law: text, cases, and materials, 3rd edn. Oxford University Press, Oxford

Bettiol L (2017) Riforma Gelli-Bianco: il ruolo delle linee guida nel giudizio di responsabilità penale in campo sanitario. Foro italiano, pp 236–241

Beumont KM (1955) *Warsaw Convention* of *1929*, as amended by the protocol signed at the Hague. J Air Law Commerce 22(4):414–433

Beveridge A (ed) (2009) Forensic investigation of explosions. Taylor & Francis, London

Bieder C, Bourrier M (eds) (2017) Trapping safety into rules: how desirable or avoidable is proceduralization? Taylor & Francis Group, Boca Raton (Florida)

Bijlsma F (2013) Justice and safety, vol 18. Hindlight, pp 62–65

Bjorge E (2015) Domestic application of the ECHR: courts as faithful trustees. Oxford University Press, Oxford

Blajev T (2013) Just culture in doubt. HindSight 18:6–7

Blanquet M (1994) L'article 5 du Traite' C.E.E. Recherce sur les obligations de fidelité des Etats membres de la Communauté. Librairie générale de droit et de jurisprudence (LGDJ), Paris

Borzì I (2014) Just culture and responsibility of air traffic controllers: some considerations regarding Judgement No. 6820/11 of the Italian criminal Court of Cassation. In: Pellegrino F (ed) Legislation and regulation of risk management in aviation activity, vol I, pp 73–98

Boysen PG (2013) Just culture: a foundation for balanced accountability and patient safety. Ochsner J 13(3):400–406

Boysen PG (2014–2015) Just culture. OT Safety Bull Focus Enforcement 7:1 ff

Bremner C (2008) Continental Airlines faces manslaughter charges over Paris Concorde crash. The Times, London

Brems E, Gerards J (eds) (2014) Shaping rightd in the ECHR: the role of the European Court of human rights in determining the scope of human rights. Cambridge University Press, Cambridge

Brooker P (2007) The European single sky needs high quality, simple incident reporting. Air Traffic Technology International, p 3 ff

Brookes A (2002) Destination disaster. Ian Allan, Birmingham, pp 22 ff

Brüggen J (2013) Why we need positive examples in our just culture. HindSight 18:44–45

Brunelli D (2014) Il "mistero" del dolo eventuale: scritti dal dibattito svoltosi a Perugia, il 27 gennaio 2012. Giappichelli, Torino

Bruni F, Tullio L (eds) (2013) Sinistri aeronautici: rischi e responsabilità. Giappichelli, Torino

Brusco C (2007) Cassazione e responsabilità penale del medico. Tipicità e determinatezza nel nuovo art. 590-sexies c.p. Diritto penale contemporaneo 11:205 ff

Brusco C (2008) Applicazioni concrete del criterio della probabilità logica nell'accertamento della causalità. Cassazione penale, pp 1875–1886

Brusco C (2018) Responsabilità medica penale: le Sezioni Unite applicano le regole sulla responsabilità civile del prestatore d'opera. Diritto penale e processo, pp 646–654

Bufo M (2012) Le competenze degli ANSPs. Il servizio informazioni volo (FIS) e il servizio informazioni volo aeroportuale (AFIS). In: Pellegrino F (ed) Air navigation rules and practices in Europe: towards harmonization. Giuffré, Milano, pp 163–190

Bufo M (2015) Gestione del rischio d'impatto tra aeromomili e volatile (c.d. Bird strike) in aeroporto e nello spazio aereo limitrofo: ripartizione delle competenze (e responsabilità) tra i principali soggetti coinvolti. In: Pellegrino F (ed) Legislation and regulation of risk management in aviation activity, vol II. Giuffré, Milano, pp 123–154

Busti S, Signorini E, Simoncini GR (eds) (2017) L'impresa aeroportuale a dieci anni dalla riforma del codice della navigazione: stato dell'arte. Giappichelli, Torino

Buzzoni A (2007) Medico e paziente. Le responsabilità civili e penali del medico e dell'équipe medica. edizioni Fag, Milano

Byers A (2002) The crash of the Concorde. The Rosen Publishing, New York

Byrne G (2002) Flight 427: anatomy of an air disaster. Copernicus Books, New York

Cagnazzo A, Toschei S, Pozzi C (eds) (2012) Le sanzioni in materia di trasporto marittimo, aereo, terrestre e Codice della strada. Giappichelli, Torino

Caldwell CL (2018) Safety culture and high-risk environments: a leadership perspective. CRC Press, Taylor & Francis, Boca Raton

Caletti GM, Mattheudakis ML (2017) Una prima lettura della legge Gelli-Bianco nella prospettiva del diritto penale. Diritto penale contemporaneo 2:85–107

Caletti GM, Mattheudakis ML (2018) La fisionomia dell'art. 590-sexies c.p. dopo le Sezioni Unite tra "nuovi" spazi di graduazione dell'imperizia e "antiche" incertezze. Diritto penale contemporaneo 4:25–46

Calleja Crespo D, Mendes de Leon P (eds) (2011) Achieving the Single European Sky: goals and challenges. Kluwer, Alphen aan den Rijn

Camarda G (2002) La responsabilità del gestore aeroportuale. Diritto dei trasporti 3:763–811

Canzio G (2015) Ragioni, verità e dubbio nel labirinto del processo penale. Giustizia penale, pp 193–198

Caputo M (2017a) La responsabilità penale dell'esercente la professione sanitaria dopo la L. n. 24 del 2017. . .quo vadis? Primi dubbi, prime risposte, secondi dubbi. Danno e responsabilità, pp 293–300

Caputo M (2017b) Promossa con riserva. La legge Gelli-Bianco passa l'esame della Cassazione e viene "rimandata a settembre" per i decreti attuativi. Rivista italiana di medicina legale, pp 724–743

Carbone V (2017) Legge Gelli: inquadramento normativo e profili generali. Corriere giuridico, pp 737–739

Carboni L (2015) Il dolo eventuale dopo la sentenza Thyssenkrupp. Key, Vicalvi (Frosinone)

Carnino A, Weimann G (eds) (1995) Proceedings of the international topical meeting on safety culture in nuclear installations. American Nuclear Society of Austria, Vienna. 24–28 April 1995

Carrol R (2010) Risk management handbook for health care organizations, 6th edn. Jossey-Bass, Chicago

Cassatella A (2017) Appeals before the European Aviation Safety Agency. In: Marchetti B (ed) Administrative remedies in the European Union: the emergence of a quasi-judicial administration. Giappichelli, Torino, pp 21–54

Castellani G (2015) Responsabilità sociale di impresa. Ragioni, azioni e reporting, Maggioli, Sant'Arcangelo di Romagna (Rimini)

Castelletta A (2004) Responsabilité médicale: droit des maladies. Dalloz, Paris

Castronuovo D (2009) La colpa peale. Giuffrè, Milano

Catalisano D (2015) The remote pilot operator in the aviation industry. Pilot perspective. In: Pellegrino F (ed) Legislation and regulation of risk management in aviation activity, vol II. Giuffré, Milano, pp 27–44

Catino M (2013) Organizational Myopia. Problems of rationality and foresight in organizations. Cambridge University Press, Cambridge

Catino M, Patriotta G (2013) Learning from errors: Cognition, emotions and just culture at the Italian Air Force. Organ Stud 34(4):437–467

Cattaneo G (1958) La responsabilità del professionista. Giuffré, Milano

Cattaneo G (1977) La responsabilità del professionista. Giuffré, Milano

Cattaneo G (1982) La responsabilità medica nel diritto italiano. In: AA.VV (ed) La responsabilità medica. Giuffré, Milano, pp 9–23

Cattaneo C, De Angelis D, Grandi M (2006) Mass disasters. In: Schmitt A, Cuhna E, Pinheiro J (eds) Forensic anthropology and medicine. Complementary sciences from recovery to cause of death. Humana Press, Totowa, pp 431–443

Cavaliere A (2017) Responsabilità medica alla luce della Riforma. Diritto.it

Cecchi R (1993) *Joint Aviation* Authorities: situazione attuale e sviluppi futuri. In: Franchi B (ed) Cinquant'anni di codice della navigazione: profili di responsabilità degli operatori del settore aereo. Aero Club d'Italia, Roma, p 129 ff

Cembrani F (2017) Su alcuni snodi critici della legge "Gelli-Bianco". Rivista italiana di medicina legale, pp 873–879

Challinor CAS (2016) ICAO Annex 13, Chicago Convention and the admissibility of air accident reports in litigation: is it time for ICAO to change course? International Institute of Air & Space Law, Montreal

Chalmers D, Davies G, Monti G (2010) European Union Law. Cambridge University Press, Cambridge, p 215 ff

Chittum S (2018) Last days of the Concorde: the crash of Flight 4590 and the end supersonic passenger travel. Smithsonian Books, Washington

Cian G, Trabucchi A (1981) Commentario breve al codice civile. Cedam, Padova

Coelho dos Santos J (2013) Just culture versus criminalization – moving forward. HindSight 18:40 ff

Cohen-Charash Y, Spector PE (2001) The *role* of justice in organizations: a meta-analysis. Organ Behav Hum Decis Process 86(2):278–321

Coman-Kund F (2018) European Union agencies as global actors: a legal study of the European Aviation Safety Agency, Frontex and Europol. Routledge/Taylor & Francis, Abingdon/New York

Comenale Pinto MM (1997) Considerazioni sull'organizzazione dell'assistenza al volo. In: Studi in onore di Gustavo Romanelli. Giuffré, Milano, p 355 ff

Comenale Pinto MM (1999) L'assistenza al volo. Evoluzione, problemi attuali, prospettive. Cedam, Padova

Comenale Pinto MM (2007) Le competenze di ENAV S.p.A. ed i servizi della navigazione aerea. In: Franchi B, Vernizzi S (eds) Il diritto aeronautico fra ricodificazione e disciplina comunitaria. Giuffré, Milano, pp 107–110

Comenale Pinto MM (2009) Responsabilità nel controllo del traffico aereo. In: Rizzo MP (ed) La gestione del traffico aereo: profili d diritto internazionale, comunitario e interno. Giuffré, Milano, pp 373–386

Comenale Pinto MM (2015a) Le raccomandazioni di sicurezza. In: Pellegrino F (ed) Legislation and regulation of risk management in aviation activity, vol II. Giuffré, Milano, pp 83–96

Comenale Pinto MM (2015b) Inchieste aeronautiche, raccomandazioni di sicurezza e "just culture". Diritto marittimo, pp 237 ff

Contissa G, Sartor G, Lanzi P, Marti P, Tomasello P (2012) Liabilities and automation in aviation. In: Proceedings SESAR Innovation Days, 27th–29th November 2012

Contissa G, Laukyte M, Sartor G, Masutti A, Lanzi P, Marti P, Tomasello P, Schebesta H (2013) Liability and automation: issues and challanges for socio-technical systems. J Airsp Oper 2:79–98

Cook A (ed) (2016) European Air Traffic Management. Principles, practice and research (safety, transparency and 'just culture'). Ashgate, Aldershot (Hampshire)

Cook R, Render M (2000) Ga*ps in the continuity* of care and progress on patient safety. Br Med J 320:791–794

Cooper MD (2000) Towards a model of safety culture. Safety Sci 36:111–136

Cooper D, Findley L (2013) Strategic safety culture roadmap. BSMS (B-Safe Management Solutions) Inc., Franklin

Cooper LC, Robertson I (eds) (1988) International review of industrial and organisational psychology. Wiley, Chichester

Cooper-Stephenson K, Gibson E (eds) (1993) Tort theory. Captus University Publications, North York

Cordero F (1957) Le situazioni soggettive nel processo penale. Giappichelli, Torino

Corrigan S (2003) Comparative analysis of safety management systems and safety culture in aircraft maintenance. Trinity College, Dublin

Costantinesco V (1987) L'art. 5 CEE. De la bonne foi à la loyauté communautaire. In: Pescatore P, Capotorti F (eds) Du droit international au droit de l'integration. Liber Amicorum Pierre Pescatore. Nomos Verlagsgesellschaft, Baden Baden, p 97 ff

Cox S, Flin R (1998) Safety culture: Philosopher's stone or man of straw? Work Stress 12 (3):189–201

Cramoisi G (2010) Air crash investigations: the end of the Concorde era. The crash of Air France Flight 4590. Mabuhay Publishing, New York

Cremona M (ed) (2018) Structural principles in EU external relations law. Bloomsbury, London

Crespi A (1955) La responsabilità penale nel trattamento medico chirurgico con esito infausto. Priulla, Palermo

Crespi A (1960) Il grado della colpa nella responsabilità professionale del medico chirurgo. Scuola positiva, pp 484 ff

Crespi A (1973) La colpa grave nell'esercizio dell'attività medico chirurgica. Rivista italiana diritto e procedura penale I:255 ff

Cromie S, Bott F (2016) Just culture's "line in the sand" is a shifting one; an empirical investigation of culpability determination. Saf Sci 86:258–272

Cromie S, Liston P, Ross D, Corrigan S, Vani L, Lynch D, Demosthenous S, Leva C, Kay A, Demosthenousb V (2013) Human and organisational factors training as a risk management strategy in an aviation maintenance company. Chem Eng 33:445–450

Croucher P (2015) Avionics in plain English. Electrocution, Calgary, pp 2–49

Cupelli C (2017a) Cronaca di un contrasto annunciato: la legge Gelli-Bianco alle Sezioni Unite. Diritto penale contemporaneo 11:244–286

Cupelli C (2017b) Il perimetro applicativo della Legge Balduzzi: aperture giurisprudenziali "vs." restrizioni normative? Processo penale e giustizia, pp 196–204

Cupelli C (2017c) La legge Gelli-Bianco nell'interpretazione delle Sezioni Unite: torna la gradazione della colpa e si riaffaccia l'art. 2236 c.c. Diritto penale contemporaneo 12:135–138

Cupelli C (2017d) La responsabilità penale degli operatori sanitari e le incerte novità della legge Gelli-Bianco. Cassazione penale, pp 1765–1778

Cupelli C (2017e) Quale (non) punibilità per l'imperizia? La Cassazione torna sull'ambito applicativo della legge Gelli. Diritto penale contemporaneo 11:250–255

Cupelli C (2018) L'art. 590-sexies c.p. nelle motivazioni delle Sezioni Unite: un'interpretazione 'costituzionalmente conforme' dell'imperizia medica (ancora) punibile. Diritto penale contemporaneo 3:246–258

Czech BA, Groff L, Strauch B (2014) Safety Cultures and Accident Investigation: Lessons Learned from a National Transportation Safety Board Forum, 13–16 October 2014, Adelaide, Australia

D'Alessandro F (2017) La responsabilità penale del sanitario alla luce della riforma "Gelli-Bianco". Diritto penale e processo, pp 573 ff

Damar D (2011) Wilful misconduct in international transport law. Springer, Hamburg

Daniels S (2017) Corporate manslaughter in the maritime and aviation industries, Lloy's Practical Shipping Guides. Informa Law from Routdlege/Taylor and Francis Group, Abingdon (Oxon)/New York

Dauer EA (2004) Ethical *misfits*: mediation and *medical* malpractice litigation. In: Sharpe VA (ed) Accountability: patient safety and policy reform. Georgetown University Press, Washington, pp 185–201

Davina A, Braithwaite J, Sandall J (eds) (2016) The sociology of healthcare safety and quality. Wiley Blackwell, Oxford, pp 31 ff

De Franchis F (1985) Law dictionary. English-Italian, vol 1. Giuffrè, Milano, p 1402

De Meo MM (2000) Riassetto del comparto aeroportuale in Italia. Gazzetta ambiente, pp 3 ff

De Simone V (1972) Sulla colpa professionale. Giustizia penale, pp 825 ff

Deiana M (2006) La definizione di disastro aviatorio nell'ordinamento penale. Diritto dei trasporti, pp 52 ff

Dekker S (2003a) Failure to adapt or adaptations that fail: contrasting models on procedures and safety. Appl Ergon 34(3):233–238

Dekker S (2003b) When human error becomes a crime. Human Factors Aerosp Saf 3(1):83–92

Dekker S (2005) The questions about human error, a new view of human factors and system safety. Taylor and Francis, London

Dekker S (2007) Just culture. Balancing safety and accountability, 1st edn. Ashgate, Aldershot (Hampshire)

Dekker S (2008) Just culture: who gets to draw the line? Cognition technology work. Springer, London

Dekker S (2010a) Balancing 'no blame' with accountability. N Engl J Med 362(3):275–276

Dekker S (2010b) Pilots, controllers and mechanics on trial: cases, concerns and countermeasures. Int J Appl Aviat Stud (IJAAS) 10(1):31–49

Dekker S (2011) The criminalization of human error in aviation and healthcare: a review. Safety Sci 49:121–127

Dekker S (2012) Just culture. Balancing safety and accountability. Ashgate, Aldershot

Dekker S (2013a) A new just culture algorithm. HindSight 18:8–9

Dekker S (2013b) Sicurezza e pensiero sistemico. La gestione della sicurezza nelle organizzazioni di oggi richiede di passare dalla caccia alle componenti rotte alla comprensione dei sistemi complessi. Hirelia ed., Milano

Dekker S (2016) Just culture: restoring trust and in your accountability organization, 2nd edn. CRC Press, Taylor & Francis, Abingdon-on-Thames

Dekker S (2016a) Just culture: restoring trust and in your accountability organization, 2nd edn. CRC Press, Taylor & Francis, Abingdon-on-Thames

Dekker S (2016b) Patient safety: a human factors approach. CRC Press, Boca Raton

Dekker SW, Breakey H (2016) 'Just culture': improving safety by achieving substantive, procedural and restorative justice. Safety Sci 85:187–193

Dekker S, Laursen T (2007) From *punitive action* to confidential reporting. Patient Saf Qual Healthc 5:50–56

Del Pinto G (2017) Il fattore umano nel controllo del traffico aereo. I quaderni di ANACNA, vol 1. Youcanprint, Roma

Delitala G (1932) Dolo eventuale e colpa cosciente. Annuario dell'Università Cattolica del Sacro Cuore, Milano

Dell'Osso AM (2016) In tema di colpa medica. Rivista italiana di medicina legale e del diritto in campo sanitario, pp 362–367

DeWinter D (2013) Experience on task in a just culture. HindSight 18:38–39

Di Bitonto ML (2017) Professione medica e procedimento penale: le novità dopo la legge n. 24/2017. Cassazione penale 10:3799–3808

Di Carlo R (2015) Dalla compliance alla performance. Possibili implicazioni dell'evoluzione delle norme di Safety Risk Manafùgement nel settore dell'aviazione civile. In: Pellegrino F (ed) Legislation and regulation of risk management in aviation activity, vol II. Giuffrè, Milano, pp 203–227

Di Giovine O (2017) Colpa penale, "legge Balduzzi" e "disegno di legge Gelli-Bianco": il matrimonio impossibile tra diritto penale e gestione del rischio clinico. Cassazione penale Cassazione penale 1:386–404

Di Landro A (2008) Interruzione del nesso causale e accertamento della causalità "modello Franzese". Foro italiano, pp 181–184

Di Palma P (2000) Il ruolo dell'ENAC per l'indirizzo, la regolazione e il controllo del settore. Gazzetta ambiente 2–3:11–12

Di Renzo (2015) Legislation and regulation of risk management in aviation activity. Civil Aviation Safety Management System. In: Pellegrino F (ed) Legislation and regulation of risk management in aviation activity, vol II. Giuffré, Milano, pp 155–162

Diederiks-Verschoor IHP, Butler MA (2006) An introduction to air law. Kluwer Law International, The Hague

Diehl A (1989) Human performance/system safety issues in aircraft accident investigation and prevention. In: Proceedings of 11th International Symposium on Aviation Psychology, Columbus, Ohio

Dismukes RK, Berman BA, Loukopoulos LD (2017) The limits of expertise. Rethinking pilot error and the causes of airline accidents. Ashgate, Farnham

Douglas M (1992) Risk and blame: essays in cultural theory. Routledge, London

Dovere S (2016) Prospettive della responsabilità penale colposa nel settore aeronautico (l'espressione 'ultra limes' della colpa lieve). Responsabilità civile e previdenza 3:1023–1038

Draetta U (2009) Elementi di diritto dell'Unione Europea. Parte istituzionale. Ordinamento e struttura dell'Unione Europea. Giuffré, Milano

Driever MJ (2002) Are evidenced-based practice and best practice the same? West J Nurs Res 24 (5):591–597

Drobnig U, David R, Egawa HH, Graveson R, Knapp V, Von Mehren AT, Noda Y, Rozmaryn S, Tschchikvadze VM, Valladão H, Yntema H (eds) (2017) International Encyclopedia of comparative law. International association of legal sciences, vol 4. Martinus Nijhoff, Brill, Leiden

Duguid B (2006) What makes negligence 'Gross' and when is misconduct 'Wilful'? blakes bulletin on energy - oil & gas. Cassels & Graydon LLP (Blakes), Toronto

Dusi G (2017) Just Culture Safety Report: atteggiamenti e comportamenti dei professionist. Youcanprint Self-Publishing, Tricase

Duval C, LeLeuch H, Pertuzio A, Weawer J (2009) International petroleum exploration and exploitation agreements: legal, economic & policy aspects, 2nd edn. Barrows Company, New York

England I (1993) The philosophy of tort law. Dartmouth Publishing Company, London

Epstein RA (1980) A theory of strict liability: toward a reformulation of tort law. Cato Institute, Washington

Erin CA, Ost S (eds) (2007) The criminal justice system and health care. Oxford University Press, Oxford

Espinola S, Costa M, Maurino D (2005) Guidance material addresses concerns about protection of safety information. ICAO J 61(6):26–28

Falcinelli D (2004) Analogia: il limite logico del sistema penale. Giurisprudenza italiana, pp 147 ff

Fanara E (ed) (2000) La nuova disciplina del trasporto aereo. Messina

Farmer A, Faure M, Vagliasindi GM (eds) (2017) Environmental crime in Europe. Bloomsbury, London

Fassert C (2001) Questioning transparency: the case of incidents. MSc thesis (mémoire de DEA). http://www.eurocontrol.int

Fassert C (2005) A comparative approach on safety in several air navigation services providers: which role for culture? ATRS (Air Transport Research Society International Conference), Rio de Janeiro, Brazil

Faure M (2010) Effective, proportional and dissuasive penalties in the implementation of the environmental crime and shipsource pollution directives: questions and challenges. Eur Energy Environ Law Rev 19(6):256–278

Federal Aviation Administration (ed) (2017) Instrument procedures handbook. Skyhorse Publishing, New York

Ferguson J, Fakelmann R (2005) The culture factor. Front Health Serv Manage 22:33 ff

Ferrara SD, Boscolo-Berto R, Viel G (eds) (2013) Malpractice and medical liability: European State of the art and guidelines. Springer, Berlin

Ferrari S (2004) Sulla valutazione della responsabilità medica per colpa. Giurisprudenza italiana 7:1492–1494

Festinger LA (1957) Theory of cognitive dissonance. Stanford University Press, Stanford

Finocchiaro M, Starrantino C (2013) The Ueberlingen case: legal scenarios after Barcelona Court of Appeal judgment. The Controller Magazine 51(4):31–33

Finucci G (1992) Riflessioni sulla responsabilità professionale del medico nella complessa situazione sanitaria moderna. Nuovo diritto, pp 420 ff

Fisher DK, Rodhes L (2013) Limitations on liability exceptions for gross negligence and willful misconduct and the implications for outsourcing agreements. Bus Technol Sourcing Rev:19

Fitzmaurice M (2016) The international convention for the prevention of pollution from ships. In: Fitzmaurice M, Martinez Gutierrez NA, Hamza R (eds) The IMLI manual on international maritime law, vol. III, Marine Environmental Law and Maritime. Oxford University Press, Oxford, pp 33–77

Fitzmaurice M, Martinez Gutierrez NA, Hamza R (eds) (2016) The IMLI manual on international maritime law, vol. III, Marine Environmental Law and Maritime. Oxford University Press, Oxford

Flin R, Mearns K, O'Conner P, Bryden R (2000) Measuring safety climate: identifying the common features. Saf Sci 34:177–192

Flin R, O'Connor P, Crichton M (2008) Safety and sharp end: a guide to non-technical skills. Ashgate, Farnham

Ford C, Jack T, Crisp V, Sandusky R (1999) Aviation accident casual analysis. Advances. Aviation Safety Conference Proceedings, pp 343 ff

Foster N (2013) Foster on EU law. Oxford University Press, Oxford

Fragali M (1967) Regolamento giuridico dell'aviazione e allegati tecnici di Chicago. Diritto aeronautico, pp 159–162

Franchi B (ed) (1993) Cinquant'anni di codice della navigazione: profili di responsabilità degli operatori del settore aereo. Aero Club d'Italia, Roma

Franchi B (2000) L'Agenzia Nazionale per la sicurezza del volo. In: Fanara E (ed) La nuova disciplina del trasporto aereo. Messina, p 595 ff

Franchi B (2006) La sicurezza del passeggero nel trasporto aereo. In: Masala L, Rosafio E (eds) Trasporto aereo e sicurezza del passeggero. Giuffrè, Milano, p 145 ff

Franchi B (2011) Controllori del traffico aereo: un'altra sentenza che ne amplia le competenze. Responsabilità civile e previdenza 11:2284–2295

Franchi B (2012) Le autorità investigative per la sicurezza dell'aviazione civile dopo l'entrata in vigore del Regolamento U.E. 996/2010. In: Pellegrino F (ed) Air navigation rules and practices in Europe: towards harmonization. Giuffré, Milano, pp 41–56

Franchi B, Vernizzi S (eds) (2007) Il diritto aeronautico fra ricodificazione e disciplina comunitaria. Giuffré, Milano

Franzoni M (2016) Colpa e linee guida. Danno e responsabilità 8/9:801–806

Freckelton I, Mendelson D (eds) (2017) Causation in law and medicine. Taaylor & Francis, Ashgate Publishing, Aldershot

Freni E (1997) La trasformazione dell'azienda di assistenza al volo in ente pubblico economico. Giornale di diritto amministrativo 7:605 ff

Furnham A (1997) The psychology of behaviour at work. Psychology Press, Hove England

Gallo M (1960) Colpa penale (diritto vigente). Enciclopedia del diritto. Giuffrè, Milano, pp 624–643

Gallo F (2001) Norme penali e norme eccezionali nell'art. 14 delle "disposizioni sulla legge in generale". Rivista di diritto civile 1:1–28

Gallo G (2002) A proposito di visual approach. In: http://www.traffico-aereo.it/pages/atcp/atcp07/atcp07.htm

Gallone G (2017) Legge Gelli: il medico "strutturato" risponde (ma, in realtà, già rispondeva) ex art. 2043 c.c. e non già ex art. 1218 c.c. Archivio giuridico della circolazione e dei sinistri stradali 5:379–387

Garland DJ, Wise JA, Hopkin VD (eds) (1999) Handbook of aviation human factors. Lawrence Erlbaum Associates Publishers, Mahwah

Garzone G, Archibald J (eds) (2010) Discourse, identities and roles in specialized communication. Peter Lang, Bern

Garzone FP, Nocco BA (2017) La responsabilità penale sanitaria nel passaggio dalla Legge Balduzzi alla Legge Gelli: "se vogliamo che tutto rimanga com'è, bisogna che tutto cambi". Rivista penale 10:885–888

Gates S (2009) Criminalisation of air accidents – time to turn the tide. Reg Int 53:9 ff

Gaur D (2005) Human factor analysis and classification system applied to civil aircraft accidents in India. Aviat Space Environ Med 76:501–505

Giannini A (1953) I trasporti aerei internazionali della convenzione di Chicago 1944. Rivista aeronautica, pp 101–103

Giannini A (1946) La convenzione di Chicago 1944 sull'aviazione civile internazionale. Rivista di diritto comm.erciale I, pp 83–92

Giannini A (1949) L'ammissione dell'Italia alla convenzione di Chicago (1944) sull'aviazione civile internazionale. Rivista di diritto della navigazione, I, pp 41–48

Giannini A (1952a) Gli emendamenti della convenzione di Chicago (1944) relativa all'aviazione civile internazionale. Rivista aeronautica, pp 729–731

Giannini A (1952b) L'organizzazione dell'ICAO. Rivista aeronautica, pp 579–584

Giannini A (1953a) Il regolamento della navigazione aerea nella convenzione di Chicago 1944. Rivista aeronautica, pp 11–16

Giannini A (1953b) La convenzione di Chicago e la sua tecnica. Rivista aeronautica, pp 269–271

Giannini A (1953c) La convenzione di Chicago 1944 sull'aviazione civile internazionale. Roma

Gibb R, Gray R, Scharff L (2016) Aviation visual perception, research and mishaps. Ashgate studies in human factors for flight operations. Taylor & Francis Group/Routdlege, London/New York

Giemulla EM, Weber L (2011) International and EU aviation law: selected issues. Kluwer, Dordrecht

Giemulla E, Schmid R, Müller-Rostin W, Dettling-Ott R, Margo R (2017) Convention for the unification of certain rules for international carriage by air. In: Giemulla E, Schmid R (eds) Montreal convention. Kluwer Law International, BV, Alphen aan den Rijn, pp 1–22

Glancey J (2015) Concorde: the rise and fall of the supersonic airliner. Atlantic Books, London

Glendon AI, Clarke SG, McKenna EF (2006) Human safety and risk management. CRC Press, Taylor & Francis, Boca Raton

Gooderham P (2009) The distinction between gross negligence and recklessness in English criminal law. J Royal Soc Med 102(9):358

Gormley LW (2017) Infringement proceedings. In: Jakab A, Kochenov D (eds) The enforcement of EU law and values: ensuring member states' compliance. Oxford University Press, Oxford. part II, para. 4, pp 65 ff

Grassi M (2007) Nesso di causalità nella responsabilità medica: note a margine di una sentenza milanese. Il Foro ambrosiano, pp 404–414

Grasso G (1979) La responsabilità penale nell'attività medico-chirurgica: orientamenti giurisprudenziali sul "grado" della colpa. Rivista italiana di medicina legale 1:80–91

Green J (2003) The *ultimate challenge for risk technologies:* controlling the accidental. In: Summerton J, Berner B (eds) Constructing risk and safety in technological practice. Routledge, London, pp 29–42

Grigoli M (1997) Profili della evoluzione normativa e sistematica dell'assistenza al volo in Italia. Trasporti (72-73):23–42

Grosse A (2010) Theroux M.P., Gross Negligence: how bad does it have to be? Bennett Jones LLP, Calgary

Guarrera G (2015) Safety Airport Management. In: Pellegrino F (ed) Legislation and regulation of risk management in aviation activity, vol II. Giuffré, Milano, pp 171–178

Guilfoyle D (2016) International criminal law. Oxford University Press, Oxford

Guldenmund FW (2000) The nature of safety culture: a review of theory and research. Saf Sci 34:215–257

Gulseth M (2007) Managing anticoagulation patients in the hospital: the inpatient anticoagulation service. American Society of Health-System Pharmacists, Bethesda, pp 17 ff

Haanappel PPC (2003) The law and policy of air space and outer space: a comparative approach. Kluwer Law International, The Hague

Hale A, Wilpert B, Freitag M (1997) After the event: from accident to organisational learning. Pergamon Press, New York

Hall HR, Roussel LA (eds) (2017) An integrative approach to research. Administration and Practice. Jones and Bartlett Learning, Burlington

Harris D (2011) Human performance on the flight deck. Ashgate, Furnham

Harris D (ed) (2013) Engineering psychology and cognitive ergonomics. Applications and services. Springer, Berlin

Harris D, Li W-C (2006) Where safety culture meets national culture: the how and why of the China Airlines CI-611 accident. Human Factors Aerosp Saf 5(4):345–353

Haslam RA, Hide SA, Gibb AGF, Gyi DE, Pavitt T, Atlinson S, Duff AR (2004) Contributing factors in construction accidents. Appl Ergon 36:401–415

Havel BF, Sanchez GS (2014) The Principles and practice of international aviation law. Cambridge University Press, Cambridge

Hawkins J (2011) Air disasters. The Rosen Publishing, New York

Hayward BJ, Lowe AR (eds) (2000) Aviation resource management. Ashgate Publishing Limited, Aldershot

Hazan M (2017) Alla vigilia di un cambiamento profondo: la riforma della responsabilità medica e della sua assicurazione. Danno e responsabilità, pp 75–91

Hazanlso M (2017) Alla vigilia di un cambiamento profondo: la riforma della responsabilità medica e della sua assicurazione (DDL Gelli). Danno e responsabilità, pp 75 ff

Heiderhoff B, Żmij G (eds) (2009) Tort law in Poland, Germany and Europe. Sellier, European Law Publishers, Munich

Heinrich HW, Petersen D, Roos N (1980) Industrial accident prevention: a safety management approach, 5th edn. McGraw-Hill, New York

Helmreich RL, Clayton Foushee H (1993) Why CRM? Empirical and theoretical bases of human factors training in aviation. In: Weiner EL, Kanki BG, Helmreich RL (eds) Cockpit resource management. Academic Press, San Diego, pp 3–57

Helreich RL, Foushee HC (1993) Why crew resource management? Empirical and theoretical bases of human factors training in aviation. In: Weiner EL, Kanki BG, Helmreich RL (eds) Cockpit resource management. Academic Press, San Diego, p 3 ff

Hempel Lindøe P, Baram M, Renn O (eds) (2014) Risk governance of offshore oil and gas operations. Cambridge University Press, Cambridge

Henley J (2001) Concorde crash 'a disaster waiting to happen'. The Guardian, 17 August 2000

Hobbs A (1997) Human factor in airline maintenance: a study of incident reports. Bureau of Air Safety investigation, Canberra (Australia)

Hobbs A, Robertson MM (1995) Human factor in airline maintenance. In: Workshop Report, Proceedings of the Third Australian *Aviation* Psychology Symposium, pp 468 ff

Hofstede G (1991) Cultures and organisational: intercultural co-operation and its importance for survival. Harper Collins, London

Hollnagel E (2004) Barriers and accident prevention. Ashgate, Aldershot

Hollnagel E (2009) The ETTO principle: efficiency-thoroughness trade-off. Ashgate, Farnham

Hollnagel E (2013) Is justice really important for safety? HindSight 18:10–13

Hondus E (2014) The development of medical liability. Cambridge University Press

Honorati C (2006) La comunitarizzazione della tutela penale e il principio di legalità nell'ordinamento comunitario. Rivista di diritto internazionale privato e processuale:941–986

Horspool M, Humphreys M (2012) European Union Law. Seventh edition core text series. Oxford University Press, Oxford

Hudson PTW (1996) Psychology and safety. Rijks Universiteit, Leiden

Hudson PTW (2001) Safety Culture: Theory and Practice. In: The Human Factor in System Reliability – Is Human Performance Predictable? (Les Facteurs humains et la fiabilité des systèmes – Les performances humaines, son telles prévisibles?). Papers presented at the Human Factors and Medicine Panel (HFM) Workshop held in Siena, Italy, from 1-2 December 1999, published in NATO series RTO (Research and Technology Organization) MP-032, p 8

Hudson PTW (2002) Real time decision making in sport. In: Moorman PP, Pipers R (eds) Proceeding Congress Sportpsychologie. Leiden University, Leiden, p 74 ff

Hudson PTW (2003a) Applying the lessons of high-risk industries to medicine. Qual Saf Health Care 12:7–12

Hudson PTW (2003b) Aviation safety culture. J Aviat Manag 4:27–48

Hudson PTW (2009) Process indicators: managing safety by the numbers. Saf Sci 47:483–485

Hudson PTW, Guchelaar H-J (2003) Risk assessment in clinical pharmacy. Pharmact World Sci 25 (3):98–103

Hudson PTW, Stephens D (2000) Cost and benefit in HSE: a model for calculation of cost-benefit using incident potential. In: Proceedings 5th SPE International Conference on Health, Safety and Environment in Oil and Gas Production and Exploration. Leiden, SPE, Richardson, Texas

Hudson PTW, Verschuur WLG, Lawton R, Parker D, Reason J (1997) Bending the rule II: the violation manual. Rijks Universiteit, Leiden

Hudson PTW, Parker D, Lawton R, van der Graaf GC (2002) Managing non-compliance: moving from theory to practice. In: Proceedings 6th SPE International Conference on Health Safety and Environment in Oil and Gas Exploration and Production, Society of Petroleum Engineers, Richardson, Texas

Humphreys L, Santos C, di Caro L, Boella G, van der Torre L, Robaldo L (2015) Mapping recitals to normative provisions. Jurix 2015: The Twenty-Eighth Annual Conference, Braga, 10-11 December 2015. In: Rotolo A (ed) Legal knowledge and information systems: proceedings. IOS Press, Amsterdam, pp 41–50. 278

Hunter DR, Baker RM (2000) Reducing accidents among general aviation pilots through a national Aviation Safety Program. In: The Fourth Australian Aviation Psychology Symposium. Aldershot, Ashgate (England)

Iacoviello FM (2014) La "Franzese": ovvero quando buone teorie producono cattiva giustizia. Critica del diritto 3:241–258

Iadecola G (*2013*) *Brevi note in tema di colpa medica dopo la c.d. legge Balduzzi. Rivista italiana di medicina legale* 1:549–553

Iadecola G (2017) Qualche riflessione sulla nuova disciplina della colpa medica per imperizia nella legge 8 marzo 2017 n. 24 (legge cd. Gelli-Bianco). Diritto penale contemporaneo 6:53–66

Infantino M, Zervogianni E (eds) (2017) Causation in European tort law. Cambridge University Press, Cambridge, pp 393 ff

Ingratoci C (2014) Notification and reporting of aircraft accidents or incidents. In: Pellegrino F (ed) Legislation and regulation of risk management in aviation activity, vol I. Giuffré, Milano, pp 293–317

Iovino A (2013) Air cops and mountain tops. HindSight 18:78–81

Ippolito F (2007) Fondamento, attuazione e controllo del principio di sussidiarietà nel diritto della Comunità e dell'Unione Europea. Giuffré, Milano

Ivaldi P (1990) Diritto uniforme dei trasporti e diritto internazionale privato. Giuffrè, Milano

Jakab A, Kochenov D (eds) (2017) The enforcement of EU law and values: ensuring Member States' compliance. Oxford University Press, Oxford

Jeffries FL (2011) Predicting safety related attitudes in the workplace: the influence of moral maturity and emotional intelligence. Institute of Behavioral and Applied Management, Anchorage (Alaska)

Jing H-S, Battea A (2016) The Dragon in the Cockpit: How Western Aviation concepts conflict with Chinese value systems. Routledge/Taylor & Francis, London/New York

Johnson CW (1995) Decision theory and safety-critical interfaces. Department of Computing Science, University of Glasgow, Glasgow, United Kingdom. Proceedings of Interact '95. In: Nordby K, Helmersen PH, Gilmore D, Johnston N (eds) Do blame and punishment have a role in organisational risk management? Chapman Hall, London, pp 121–127

Johnson C (2005) Visualizing the relationship between human error and organizational failure. Department of Computing Science, University of Glasgow, Glasgow. http://www.dcs.gla.ac.uk/

Johnson CW (2008) Have we learned enough from Überlingen: the challenges of safety improvement in European Air Traffic Management, University of Glasgow, Glasgow. https://pdfs. semanticscholar.org/5856/b38ce8b127ee974f19ec6f7b8415c2cc67db.pdf and www.dcs.gla.ac. uk/~johnson

Johnston N, McDonald N, Fuller R (eds) (1994) Applications of psychology to the aviation system. Averbury Aviation, Aldershot

Kaminski-Morrow D, Learmount D (2004) Spate of errors led to Linate crash. Flight International, London

Katri N, Brown GD, Hicks LL (2009) From a blame culture to a just culture in health care. Health Care Manag Rev 34(4):312–322

Kennedy R, Kirwan B (1995) The failure mechanisms of safety culture. In: Carnino A, Weimann G (eds) Proceedings of the international topical meeting on safety culture in nuclear installations, American Nuclear Society of Austria, Vienna, 24–28 April 1995, 27(13):281–290

Khatib TM (1996) Organizational culture, subcultures, and organizational commitment. Iowa State University, Ames. https://lib.dr.iastate.edu/rtd

Kim K (1989) Human reliability model with probabilistic learning in continuous time domain. Microelectron Reliab 29(5):301–311

Klimas T, Vaiciukaite J (2008) The law of recitals in European community legislation. ILSA J Int Comp Law 15:1–33

Koch BA (ed) (2011) Medical liability in Europe: a comparison of selected jurisdictions. De Gruyter, Vienna

Kodate N, Dodds A (2008) Factors affecting willingness to report patient safety incidents in hospitals. NIHR King's Patient Safety and Service Quality Centre (PSSQ) Risk Programme, Working paper 1, Cambridge University Press, Cambridge, pp 2–38

Kohn LT, Corrigan JM, Donaldson MS (eds) (2000) To *Err* is human: building a safer health system. Institute of Medicine (US), Committee on Quality of Health Care in America, National Academies Press, Washington DC, pp 1–8

Koivu H (2013) Just and safety: the art of making mistakes. People sell washing machines – robots fly aeroplanes? Hindsight 18:66–69

La Penna V (1992) L'Azienda nazionale di assistenza al volo: ente pubblico o azienda autonoma? Rivista della Corte dei conti 4:182–195

La Penna V (1993) L'Azienda nazionale di assistenza al volo: ente pubblico o azienda autonoma? in Enti pubblici, pp 379 ff

La Torre U (2015a) Aircraft pilotage. Ordines 1:2–28

La Torre U (2015b) Aircraft pilotage. In: Pellegrino F (ed) Legislation and regulation of risk management in aviation activity, vol II. Giuffré, Milano, pp 3–25

La Torre U (2016) Il "visual approach" notturno nella giurisprudenza italiana. Rivista diritto navigazione 1:75–105

La Torre U, Moschella G, Pellegrino F, Rizzo MP, Vermiglio G (eds) (2006) Studi in memoria di Elio Fanara, vol 1. Giuffré, Milano

Larick J (2018) Pars pro toto: the Member States' obligations of sincere cooperation, solidarity and unity. In: Cremona M (ed) Structural principles in EU external relations law. Bloomsbury, London

Laudan L (2006) *Truth, error* and criminal law: an essay in legal epistemology. Cambridge University Press, Cambridge

Lawrenson AJ, Braithwaite GR (2018) Regulation or criminalisation: what determines legal standards of safety culture in commercial aviation. Safety Sci 102:251–262

Lawson FH, Markesinis BF (1982) Tortious liability for unintentional harm in the common law and the civil law, vol I. Cambridge University Press, Cambridge

Leape L (1994) Error in medicine. J Am Med Assoc 272(23):1851–1857

Leplat J (1982) Accidents and incidents production: methods of analysis. J Occup Accid 4 (2–4):299–310

Lerner JS, Tetlock PE (1999) *Accounting* for the effects of *accountability*. Psychol Bull 125 (2):255–275

LeSage P, Dyar JT, Evans B (2011) Crew resource management: principles and practice, developing a culture for open communication. Jones and Bartlett Publishers, Sudbury

Li WC, Harris D (2005) HFACS analysis of ROC air force aviation: reliability analysis and cross-cultural comparison. Int J Appl Aviat Stud 5:65–81

Lichfield J (2010) Air France grounds Concorde until cause of crash is known. The Independent, London, 18 October 2010

Licu T, van Dam R (2013) Just culture in aviation: dynamics and deliverables. HindSight 18:18–21

Licu T, van Dam R (2014) Just culture in aviation: dynamics and deliverables. In: Pellegrino F
(ed) Legislation and regulation of risk management in aviation activity, vol I. Giuffré, Milano,
pp 55–65

Licu T, Baumgartner M, van Dam R (2013) Everything you always wanted to know about just
culture (but we afraid to ask). HindSight 18:14–17

Lintner T, Dunlap T (2013) Just culture and American jurisprudence. Hinsight 18:52–55

Liu H, Zhang C (2007) Study on safety comprehensive assessment of construction site in man-
machine-environment system. J Chongqing Jianzhu Univ, pp 107–111

Lobato de Faria P (2010) Medical law in Portugal. Kluwer, Alphen aan den Rijn, pp 99 ff

Long S, Dhillon BS (eds) (2016) Man-machine-environment system engineering. In: Proceedings
of the 17th International Conference on MMESE, Springer, Singapore

Lu AC-J (2013) International airline alliances: EC competition law/US. Kluwer Law International,
The Hague

Ludwig DA (2009) Safety management systems for airports, vol 2. Guidebook, Transportation
Research Board, Washington

Luparia L (2002) Obbligatorietà e discrezionalità dell'azione penale nel quadro comparativo
europeo. Giurisprudenza italiana 8:1751–1758

Lyng S (1990) Edgework: a social psychological analysis of voluntary risk taking. Am J Social 95
(4):851–856

Magnosi S (2010) Il nuovo assetto organizzativo dell'Ente Nazionale per l'Aviazione Civile.
Rivista di Diritto della navigazione, pp 313–318

Mandel DR (2003) Effect of counterfactual and factual thinking on causal judgements. Think
Reason 9(3):245–265

Manners-Bell J (2014) Supply chain risk: understanding emerging threats to global supply chains.
Kogan Page, London

Marais K, Dulac N, Leveson N (2004) Beyond normal accidents and high reliability organizations:
the need for an alternative approach to safety in complex systems. Engineering Systems
Division Symposium, MIT, Cambridge, pp 1–16

Marandola A (2012) Ricostruzione "alternativa" del fatto e test di ragionevolezza del "dubbio" in
appello. Archivio penale, pp 365–374

Marchetti B (ed) (2017) Administrative remedies in the European Union: the emergence of a quasi-
judicial administration. Giappichelli, Torino

Marino A (2013) Agenzie e Autorità di regolazione del trasporto nel diritto comunitario e interno.
Jovene, Napoli

Marseglia G, Viola L (2007) La responsabilità penale e civile del medico, Halley ed., Matelica

Marti P, Lanzi P, Bannon L, Sartor G, Contissa G, Masutti A (2011) Liability and Automatation:
issues and challenges for socio-technical systems. In: Proceedings SESAR Innovation Days,
29 November-1 December 2011, SESAR WPE, pp 1-5

Martin J (2013) Key facts criminal law, 4th edn. Routledge, Abingdom, p 79 ff

Martin J, Storey T (2015) Unlocking criminal law, 5th edn. Routledge, Abingdon

Martorell S, Guedes Soares C, Barnett J (eds) (2014) Risk analysis: theory, methods and applications. CRC Press, Taylor and Francis, Boca Raton

Marx D (1997) Discipline: the role of rule violation. Ground Eff 2:1–7

Marx D (2001) Patient safety and the 'just culture': a primer for health care executives. Columbia
University, New York

Marx D (2009) Whack-a-mole: the price we pay for expecting perfection. Plano, Texas

Marx D (2012) Just Culture Algorithm™ V3. 2 – for employers. The Just Culture community.
Outcome Engenuity, Eden Prairie

Masala L, Rosafio E (eds) (2006) Trasporto aereo e sicurezza del passeggero. Giuffrè, Milano

Masera L (2007) Accertamento alternativo ed evidenza epidemiologica nel diritto penale. Giuffré,
Milano, p 166 ff

Masieri C (2016) Accertamento del nesso di causalità nella responsabilità sanitaria: paradigma
penale e civile a confronto. Rivista trimestrale diritto processuale civile, pp 1439–1462

Mason K (2017) Fault, causation and responsibility: is tort law just an instrument of corrective justice? In: Freckelton I, Mendelson D (eds) Causation in law and medicine. Taylor & Francis, Ashgate Publishing, Aldershot, pp 147 ff

Massaro A (2017) La legge Balduzzi e la legge Gelli-Bianco sul banco di prova delle questioni di diritto intertemporale alle Sezioni unite l'ardua sentenza. Giurisprudenza penale web, pp 1–8

Masutti A (2010) Prospettive di riforma dell'aviazione civile. Diritto dei trasporti, pp 473–483

Masys A (ed) (2016a) Disaster forensics: understanding root cause and complex causality. Springer, Ottawa

Masys A (2016b) Patient safety and disaster forensics: understanding complex causality through actor network ethnography. In: Masys A (ed) Disaster forensics: understanding root cause and complex causality. Springer, Ottawa, pp 63–82

Mateesco Matte N (1995) La Convenzione di Chicago. Quo vadis OACI? In: Studi in onore di Antonio Lefebvre d'Ovidio, Giuffré, Milano, pp 641 ff

Mauceri F (2015) Al di là di ogni ragionevole dubbio o più probabile che non: note minime sul nesso causale nella responsabilità civile. Jus civile, pp 110–116

McDonald N, Corrigan S, Daly C, Cromie S (2000a) Safety management systems and safety culture in aircraft maintenance organisations. Saf Sci 34:151–176

McDonald N, Corrigan S, Cromie S, Daly C (2000b) An organizational approach to human factors. In: Hayward BJ, Lowe AR (eds) Aviation resource management. Ashgate Publishing Limited, Aldershot, pp 51–61

McDonald N, Corrigan S, Ward M (2002a) Cultural and organizational factors in system safety: good people in bad systems. In: Proceedings of the 2002 International Conference on Human-Computer Interaction in Aeronautics, HCI-Aero, pp 205–209

McDonald N, Corrigan S, Ward M (2002b) *Well-intentioned* people in dysfunctional systems, Keynote presented at Fifth Workshop on Human Error, Safety and Systems Development, Newcastle

McGrath J (2015) The Little book of big decision models. Pearson Education Limited, Birmingham City University, Birmingham

Mearns K, Whitaker SM, Flin R (2003) Safety climate, safety management practice and safety performance in offshore environments. Saf Sci 41:641–680

Merry AF (2009) How does the law recognize and deal with medical errors? J Roy Soc Med 102:265–271

Merry AF, McCall Smith A (2001) Errors, medicine and the law. Cambridge University Press, Cambridge

Michaelides-Mateou S, Mateou A (2012) Flying in the face of criminalization: the safety implications of prosecuting aviation professionals for accidents. Routledge, Abingdon-on-Thames

Michaelides-Mateou S, Mateou A (2015) Notification and reporting of aircraft accidents or incidents. In: Pellegrino F (ed) Legislation and regulation of risk management in aviation activity, vol I. Giuffré, Milano, pp 265 ff

Miettinen S (2013) Criminal law and policy in the European Union. Taylor & Francis, Abingdon

Miglio M (2016) In tema di responsabilità medica colposa. Rivista italiana medicina legale e diritto in campo sanitario 3:1250–1256

Milde M (2008) International air law and ICAO. Eleven International Publishing, Utrecht

Milde M (2012) International air law and ICAO. In: Benkö M (ed) Essential air and space law, vol 10. Elven Law, Utrecht, pp 8–10

Miller VB, Jones TL (2011) Creating a just culture: a nurse leader's guide. Hcpro Incorporated, Middleton

Moccheggiani M (2017) Sapere scientifico e ruolo del giudice. Primi appunti. Quaderni costituzionali 3:571–597

Monaco R (1981) Le funzioni dell'OACI. In: Monaco R (ed) Scritti di diritto delle organizzazioni internazionali. Giuffré, Milano, p 403 ff

Moorman PP, Pipers R (eds) (2002) Proceeding Congress Sportpsychologie. Leiden University, Leiden

Morley FJ, Harris D (2006) Ripples in a pond: an open system model of the evolution of *safety culture*. Int J Occup Saf Ergon 12(1):3–15

Morreim H (2004) *Medical errors: pinning* the blame versus blaming the system. In: Sharpe VA (ed) Accountability patient safety and policy reform. Georgetown University Press, Washington, pp 213–232

Morselli E (1990) Analogia e fattispecie penale. L'indice penale, pp 505 ff

Mortelmans K (1998) The *principle* of loyalty to the community (Article 5 EC) and the obligations of the community institutions. Maastricht J Eur Comp Law 5:67–88

Muchinsky PM (1997) Psychology applied to work, 5th edn. Brooks/Cole, Pacific Grove (California), United States

Mura A (2017) Attuale insostenibilità dell'epistemologia sottesa alla sentenza Franzese. Cassazione penale 9:3396–3413

Murray SR (1997) Deliberate decision making by aircraft pilots: a simple reminder to avoid decision making under panic. Int J Aviat Psychol 7:83–100

Muscolo P (1970) Fondamento, natura e limiti della colpa medica. Giustizia penale 11:449–468

Nastro V (1993) Assistenza al volo e controllo del traffico aereo. Hoepli, Milano

Nefeli Gribaudi M (2012) Consenso e dissenso informati nella prestazione medica. Giuffrè, Milano

Nemsick J, Passeri SG (2012) Criminalizing aviation: placing blame before safety. American Bar Association Journal, pp 21 ff

Nizzo C (1997) *L'art. 5 del Trattato CE* e la clausola di buona fede nell'integrazione europea. Diritto dell'Unione Europea:381–392

Nocco L (2006) Causalità: dalla probabilità logica (nuovamente) alla probabilità statistica, la Cassazione civile fa retromarcia. Danno e responsabilità 12:1239–1245

Nordby K, Helmersen PH, Gilmore D, Johnston N (1995) Do blame and punishment have a role in organisational risk management? Flight Deck 15:33–36

Norman DA (1988) The psychology of everyday things. Basic Book, New York

Norman DA (2013) The design of everyday things. Basic Book, New York

North DM (2000) Let judicial system run its course in crash cases. Aviat Week Space Technol 152 (20):66–67

Nuvolone P (1969a) Colpa civile e colpa penale. In: Nuvolone P (ed) Trent'anni di diritto e procedura penale. Cedam, Padova

Nuvolone P (1969b) Trent'anni di diritto e procedura penale. Cedam, Padova

O'Leary M, Chappel SL (1997) Confidential incident reporting systems create vital awareness of safety problems. ICAO J 51:11–13

O'Neill A (2001) EU law for lawyer. Hart Publishing, Oxford

Oberdiek H (1972) Intention and foresight in criminal law. Mind 81(323):389–400

Orasanu JM (1993) Decision-making in the cockpit. In: Weiner EL, Kanki BG, Helmreich RL (eds) Cockpit resource management. Academic Press, San Diego, pp 137–172

Ormerod D (2005) Smith & Hogan's criminal law, 11th edn. Oxford University Press, Oxford

Ormerod D, Laird K (2014) Text, cases, and materials on criminal law. Oxford University Press, Oxford

Oxford Aviation Academy (2008) Air law, ATPL ground training series. Oxford Aviation Academy, Oxford, p 145

Page AH (2007) Making just culture a reality: one organization's approach. Agency for Healthcare Research and Quality – AHRQ.gov, in https://psnet.ahrq.gov

Paliero CE (2015) Causalità e probabilità tra diritto penale e medicina legale. Rivista italiana di medicina legale e del diritto in campo sanitario 4:1507–1518

Palma A (2017) Molto rumore per nulla: la Legge Gelli-Bianco di riforma della responsabilità penale del medico. Rivista italiana di medicina legale e diritto in campo sanitario 2:523–541

Palmieri A (2018) In tema di responsabilità colposa per morte o lesioni personali in ambito sanitario. Foro italiano 4:235–238

Palombi E (2018a) Il rispetto del diritto vivente. Archivio giuridico circolazione assicurazuine e responsabilità 4:312–316

Palombi E (2018b) Il rispetto del diritto vivente. Rivista penale 4:352–355

Palombi E (2018c) Profili penali della colpa medica nell'evoluzione giurisprudenziale. Archivio giuridico circolazione assicurazione e responsabilità 1:32–39

Pandit MS (2009) Medical negligence: criminal prosecution of medical professionals, importance of medical evidence: some guidelines for medical practitioners. Indian J Urol 25(3):379–383

Panelli S (2015) Inchiesta aeronautica e procedimento penale. In: Pellegrino F (ed) Legislation and regulation of risk management in aviation activity, vol II. Giuffré, Milano, pp 231–237

Panelli S, Scarabello M (2013) Why is it necessary to criminalise negligent behaviour? HindSight 18:58–61

Pardolesi R, Simone R (2017) Nuova responsabilità medica: il dito e la luna (contro i guasti da contatto sociale?). Foro italiano 4:161–175

Parker-Holland J (2018) A recipe for disaster? Gross negligence manslaughter in R vs. Zaman. In: Keep Calm and Talk Law, www.keepcalmtalklaw.co.uk

Patankar MS, Brown JP, Sabin EJ, Bigda-Peyton TG (2012) Safety culture: building and sustaining a cultural change in aviation and healthcare. Saint Louis University (USA), Ashgate, Farnham Surrey, England

Pavich G (2017) La responsabilità penale dell'esercente la professione sanitaria: cosa cambia con la legge Gelli-Bianco. Cassazione penale 7–8:2961–2978

Pellegrino F (2003) Sull'applicabilità dell'Annesso 13 ICAO nell'ordinamento italiano. Diritto dei trasporti, pp 805 ff

Pellegrino ED (2004) Prevention and medical error: where professional and organizational ethics meet. In: Sharpe VA (ed) Accountability patient safety and policy reform. Georgetown University Press, Washington, pp 83–98

Pellegrino F (2007) Sicurezza e prevenzione degli incidenti aeronautici nella normativa internazionale, comunitaria e interna. Giuffré, Milano

Pellegrino F (2009) La Direttiva 2006/23/CE in materia di licenza comunitaria dei controllori di volo. In: Rizzo MP (ed) La gestione del traffico aereo: profili di diritto internazionale, comunitario ed interno, vol 242. Giuffrè, Milano, pp 211–232

Pellegrino F (ed) (2012a) Air navigation rules and practices in Europe: towards harmonization. Giuffré, Milano

Pellegrino F (2012b) Il contratto di trasporto aereo e la sanzione amministrativa. In: Cagnazzo A, Toschei S, Pozzi C (eds) Le sanzioni in materia di trasporto marittimo, aereo, terrestre e Codice della strada, vol II. Giappichelli, Torino, pp 328–352

Pellegrino F (2013a) Il concetto di just culture nel diritto aeronautico. In: Turco Bulgherini E, Salerno F (eds) Infrastrutture e navigazione: nuovi profili della sicurezza marittima ed aerea. Aracne, Roma, pp 129–146

Pellegrino F (2013b) Just culture principles in Aviation law from a European perspective. Ann Air Space Law, pp 471–490

Pellegrino F (ed) (2014) Legislation and regulation of risk management in aviation activity, vol I. Giuffré, Milano

Pellegrino F (ed) (2015) Legislation and regulation of risk management in aviation activity, vol II. Giuffré, Milano

Pepe J, Cataldo PJ (2011) Manage risk, build a just culture. Health Prog 92(4):56–60

Perez Gonzales JD (2012) ICAO: human factors, management and organization. Aeroscience, pp 13–16

Perleth M, Jakubowski E, Busse R (2001) What is 'best practice' in health care? State of the art and perspectives in improving the effectiveness and efficiency of the European health care systems. Health Policy 56(3):237–238

Perrow C (1984) Normal accidents: living with high risk technologies. Basic Books, New York

Pescatore P, Capotorti F (eds) (1987) Du droit international au droit de l'integration. Liber Amicorum Pierre Pescatore. Nomos Verlagsgesellschaft, Baden Baden

Petrick-Felber N (2014) Liberalizing Europe's Skies – a failure? An analysis of airline entry and exit in the post-liberalized German Airline Market, 1993–2006. Anchor Academic Publisher, Hamburg

Petschonek SL (2011) Developing the just culture assessment tool: a method for measuring individual cultural perceptions in a healthcare setting. University of Memphis, Memphis

Piao CS, Cheng WM, Zhou G (2009) Safety analysis and assessment for human-machine-environment systematic engineering. Ind Saf Environ Prot:44–46

Pidgeon N (1998) Safety culture: key theoretical issues. Work Stress 12(3):202–216

Pidgeon N, O'Leary M (1994) Organisational safety culture: implications for aviation. In: Johnston N, McDonald N, Fuller R (eds) Applications of psychology to the aviation system, Averbury Aviation. Aldershot, *Hampshire*, pp 21–43

Pidgeon N, O'Leary M (2000) Man-made disasters: why technology and organizations (sometimes) fail. Saf Sci 34:15–30

Pifisterer H (2017) European regulation of aerodrome safety management system in the EASA system. Kassel University

Piqué F (2017) Il "dolo colpito a mezza via dall'errore": un terreno ancora scivoloso per l'interprete, dove si scontrano esigenze di giustizia sostanziale e necessità di rigorosa applicazione dei presidi garantistici vigenti nel nostro ordinamento. Cassazione penale, pp 1083–1099

Piras P (2018) Un distillato di nomofilachia: l'imperizia lieve intrinseca quale causa di non punibilità del medico. Diritto penale contemporaneo 5:1–11

Pizzi C (2015) The Franzese Judgment: a look at open problems. Paper read at the Conference "Casualty, Counterfactuals and Legal Responsibility", Sassari, 8 October 2015, pp 1 ff. http://www.academia.edu/

Polidori R (2014) Prova indiziaria e giudizio di colpevolezza "oltre ogni ragionevole dubbio". Diritto penale e processo 5:574–585

Pontonio F, Pontonio C (2006) In tema di responsabilità penale del medico: l'art. 2236 c.c. non trova diretta applicazione. Ginecologia Ostetricia Forense 1(1):31–36

Ponzanelli G (2017) Medical malpractice': la legge Bianco-Gelli. Contratto e impresa 2:356–363

Pooley E (2013) The accident investigation. HindSight 18:71–73

Preti F (2012) Il ruolo del Performance Review Body nel Cielo Unico Europeo: la regolazione delle prestazioni nell'Unione europea. In: Pellegrino F (ed) Air navigation rules and practices in Europe: towards harmonization. Giuffré, Milano, pp 113–119

Prosdocimi S (1993) Dolus eventualis: il dolo eventuale nella struttura delle fattispecie penali. Giuffré, Milano

Pugliese SAF (1989) Competenze civili e militari nell'esercizio dei servizi di assistenza al volo. Diritto e pratica dell'aviazione civile:232 ff

Quercietto L (1954) Modifiche alla convenzione di Chicago decise dall'Assemblea dell'O.A.C.I. Rivista aeronautica, pp 1225–1231

Rallo N (2014) ICAO Annex 19: what implications for the state safety programme and for safety oversight? In: Pellegrino F (ed) Legislation and regulation of risk management in aviation activity, vol I. Giuffrè, Milano, pp 101–107

Randazzo V (2004) Alcuni profili problematici relativi all'attribuzione di funzioni all'Agenzia europea per la sicurezza aerea. Diritto dell'Unione Europea 4:847–867

Rasmussen J (1982) Human errors. A taxonomy for describing human malfunction in industrial installations. J Occup Accid 4(2–4):311–333

Rasmussen J (1983) Skills, rules and knowledge: signals, signs and symbols and other distinctions in human performance models. IEEE Trans Sys Man Cybern SMC-13(3):257–266

Rasmussen J, Duncan K, Leplat J (eds) (1987) New technology and human error. John Wiley and Sons, New York

Reason J (1990a) Human error. Cambridge University Press, Cambridge

Reason J (1990b) The contribution of latent human failures to the breakdown of complex systems. Philos Trans Roy Soc Series B Biol Sci, pp 475–484

Reason J (1995a) A systems approach to organizational error. Ergonomics 38(8):1708–1721

Reason J (1995b) Comprehensive error management in aircraft engineering: a manager's guide. British Airways Engineering, London

Reason J (1997) Managing the risks of organizational accidents. Ashgate, Aldershot, e-book published in 2016

Reason J (1998) Achieving a safe culture: theory and practice. Work Stress 3:293–306

Reason J (2000a) Human error: models and management. Br Med J, pp 768–770

Reason J (2000b) Safety paradoxes and safety culture. Injury Control Saf Promot 7(1):3–14

Reason J, Parker D, Free R (1994) Bending the rules: the varieties, origins and management of safety violations. Rijks Universiteit, Leiden

Reiman T, Pietikäinen E (2010) Indicators of safety culture – selection and utilization of leading safety performance indicators. Report number: 2010:07

Reinhart RO (1996) Basic flight physiology, 2nd edn. Mcgraw-Hill, New York

Reuter P (2013) Just culture in the real world: flight safety and the realities of society. HindSight 18:49–51

Riguzzi M (1974) In tema di colpa temeraria e consapevole prevista dall'art. XIII del Protocollo dell'Aja del 1955. Diritto aereo:246–264 ff

Risicato L (2013) L'attività medica di équipe tra affidamento ed obblighi di controllo reciproco. L'obbligo di vigilare come regola cautelare. Giappichelli, Torino

Risicato L (2018) Le Sezioni unite salvano la rilevanza "in bonam partem" dell'imperizia "lieve" del medico. Giurisprudenza italiana 4:948–954

Rizzo MP (2006) Il pacchetto di regolamenti comunitari per la realizzazione del "cielo unico europeo". In: La Torre U, Moschella G, Pellegrino F, Rizzo MP, Vermiglio G (eds) Studi in memoria di Elio Fanara, vol 1. Giuffrè, Milano, pp 407–441

Rizzo MP (ed) (2009) La gestione del traffico aereo: profili d diritto internazionale, comunitaria e nazionale. Giuffré, Milano

Romanelli G (1994a) Assistenza al volo: da attività di polizia della navigazione a prestazione di servizio. In: Romanelli G (ed) Spunti di studio sull'attività di assistenza al volo, Cagliari, pp 5 ff

Romanelli G (ed) (1994b) Spunti di studio sull'attività di assistenza al volo, Cagliari

Roncali D (2017) Le linee guida e le buone pratiche: riflessioni medico-legali a margine della legge Gelli-Bianco. Danno e responsabilità 3:280–282

Rose D (2001) Concorde: the unanswered question. The Guardian, 13 May 2001

Rossi LS (2018) 2, 4, 6 (TUE)… l'interpretazione dell'"Identity Clause" alla luce dei valori fondamentali dell'UE. In: Codinanzi M, Cannizzaro V, Adam R (eds) Liber Amicorum in onore di Antonio Tizzano: De la Cour CECA à la Cour de l'Union: le long percours de la justice européenne. Giappichelli, Torino, pp 859–870

Rossi dal Pozzo F (2014) EU legal framework for safeguarding air passenger rights. Springer, Heidelberg

Roughton J, Mercurio J (2002) Developing an effective safety culture: a leadership approach. Butterworth-Heinemann, Woburn

Rousseau D (1988) Quantitative Assessment of Organisational Culture: the case for multiple measures. In: Cooper LC, Robertson I (eds) International review of industrial and organisational psychology. Wiley, Chichester, pp 139–158

Ruitenberg B (2002) Court case against Dutch controllers. The Controller, p 22 ff

Salaman G (ed) (1992) Human resource strategies. Sage, London

Salinas de Frias AM (2013) Counter-terrorism and human rights in the case of the European Court of Human Rights. Council of Europe, Strasbourg

Saponaro L (2018) Il dubbio ragionevole alla ricerca di una definizione. Giurisprudenza italiana 2:469–473

Sardella B (1995) "Eurocontrol" non è un'impresa secondo le norme del diritto comunitario. Giustizia civile 1:12–16

Scala M (2013) If it had happened in your country, what would the judgement have been? HindSight 18:80–81

Scarabello M (2017) Work-as-imagined. Work-as-done, and the rule of law. HindSight 25:40–41

Schein EH (2010) Organizational culture and leadership, 4th edn. Jossey-Bass, San Francisco

Schmitt A, Cuhna E, Pinheiro J (eds) (2006) Forensic anthropology and medicine. Complementary sciences from recovery to cause of death. Humana Press, Totowa

Scholtze M (2014) Gestione e mitigazione del rischio nelle attività produttive complesse – Il sistema aviazione. In: Bruni F, Tullio L (eds) Sinistri aeronautici: rischi e responsabilità. Giappichelli, Torino

Schubert F (2013) A just culture in aviation – who is an expert? HindSight 18:46–48

Schültze R, Tridimas T (eds) (2018) Oxford principles of Europeam Union law. Oxford University Press, Oxford

Schwenk W, Schwenk R (1998) Aspects of international cooperation in air traffic management. Martinus Nijhoff, The Hague, p 129 ff

Sciolla Lagrange A (1990) Organizzazione dell'aviazione civile internazionale (OACI). Enciclopedia Giuridica Treccani, XXXI, p 1 ff

Scognamiglio C (2017) Regole di condotta, modelli di responsabilità e risarcimento del danno nella nuova legge sulla responsabilità sanitaria. Corriere giuridico 34(6):740–748

Senders JW, Moray NP (1991) Human error: cause, prediction, and reduction. Lawrence Erlbaum Associates, Hillsdale

Serrao F (1989) L'avvio dell'attività dell'Azienda autonoma di assistenza al volo per il traffico aereo generale. Enti pubblici, 863 ff

Shappell SA (2000) The Human Factors. Analysis and Classification System-HFACS, DOT/FAA/AM-00/7

Shappell SA, Wiegmann D (1997) A human error approach to accident investigation: the taxonomy of unsafe operations. Int J Aviat Psychol 7:269–291

Shappell S, Wiegmann D (2001) Applying reason: the human factors analysis and classification system. Human Factors Aerosp Saf 1(1):59–86

Sharpe CC (1999) Nursing malpractice: liability and risk management. Greenwood publishing Group, Westport

Sharpe VA (2003) *Promoting patient safety*: an ethical basis for policy deliberation. Hast Center Rep Spec Suppl 33(5):3–18

Sharpe VA (ed) (2004) Accountability patient safety and policy reform. Georgetown University Press, Washington

Sharpe VA (ed) (2015) Accountability: patient safety and policy reform. Georgetown University Press, Washington

Shorrock S (2013) 'Human error'. The handicap of human factors. HindSight 18:32–37

Simpson P (2014) The Mammoth Book of air disasters and near misses. Running Press Book Publishers, Philadelphia

Sinay J (2014) Safety management in a competitive business environment. CRC Press/Taylor & Francis Group, Boca Raton/Florida

Sirena P (2013) Il settore aeronautico. L'elaborazione giurisprudenziale italiana. Relazione all'Incontro di Studio organizzato dalla Scuola Superiore della Magistratura, Centro Alti Studi Difesa: "La responsabilità colposa, posizioni di garanzia e profili concausali in attività complesse di organizzazioni a rischio consentito. Due esperienze a confronto, il settore aeronautico e quello sanitario", Roma, 20 novembre 2013

Sirena PA (2014) The case law of the Italian Supreme Court of Cassation in the field of aviation accident. Eurocontrol Seminar on "Aviation and the Judiciary", 30 October 2014, Bruxelles

Sirena PA (2015a) The case law of the Italian Supreme Court of Cassation in the field of aviation accidents. In: Pellegrino F (ed) Legislation and regulation of risk management in aviation activity, vol II. Giuffré, Milano, pp 252–260

Sirena P (2015b) The Italian Supreme Court of Cassation. A case study of Aviatin Accident. The Controller The Controller Magazine, IFATCA, pp 6 ff

Skegg P (1998) *Criminal prosecutions* of negligent health professionals: the New Zealand experience. Med Law Rev 6(2):220–246

Smith D (2001) Controlling Pilot Error, Controlled Flight Into Terrain (CFIT/CFTT). McGraw-Hill Education, Milano

Somma E (2014) "Oltre ogni ragionevole dubbio". Una formula enfatica da contestualizzare: meglio, da evitare. Rivista italiana di diritto e procedura penale 1:366–373

Song X, Xie Z (2014) Application of man-machine-environment system engineering in coal mines safety management. In: "2014 ISSST", 2014 International Symposium on Safety Science and Technology, Procedia Engineering 84:87–92

Spadoni AS (2001) Eurocontrol e la funzione unificatrice dello spazio aereo europeo: rafforzamento o svuotamento del principio della sovranità degli Stati. Rivista di diritto pubblico e scienze politiche 2:259–267

Sparck-Jones K (1972) A statistical interpretation of term specificity and its application in retrieval. J Doc 28(1):11–21

Spitzmiller R (2011) Selected areas of Italian tort law: cases and materials in a comparative perspective. Fagnano Alto (L'Aquila)

Stadler G (2009) The role of eurocontrol in the implementation of the Single European Sky. In: Rizzo MP (ed) La gestione del traffico aereo: profili di diritto internazionale, comunitario ed interno. Giuffrè, Milan, pp 267–279

Starrantino C, Finocchiaro M (2013) The judicial aftermath. HindSight 18:74–77

Starrantino C, Finocchiaro M (2014) Just culture. A "new" approach to safety. In: Pellegrino F (ed) Legislation and regulation of risk management in aviation activity, vol I, pp 23–54

Steel N, Abdelhamid A, Stokes T, Edwards H, Fleetcroft R, Howe A, Qureshi N (2014) A review of clinical practice guidelines found that they were often based on evidence of uncertain relevance to primary care patients. J Clin Epidemiol 67(11):1251–1257

Sternberg R (1996) Cognitive psychology, 2nd edn. Harcourt Brace College Publishers, San Diego

Stolzer AJ, Goglia JJ (eds) (2015) Safety management systems in aviation, 2nd edn. Routledge, Taylor & Francis Group, Abingdon, Oxson

Stolzer AJ, Halford CD, Goglia JJ (2011) Implementing safety management systems in aviation. Ashgate, Farnham (England)

Strachan DMA (1970) The scope and application of the "but for" causal test. Med Legal Regul (MLR) 33:386–389

Strauch B (2004) Investigating human error: incidents, accidents, and complex systems. Ashgate, Aldershot

Strauch B (2017) Investigating human error: incidents, accidents, and complex systems. CRC Press, Taylor & Francis, Boca Raton

Stravino P (2012) Quali sono i rapporti tra il delitto di cui all'art. 335 c.p. e l'illecito amministrativo di cui al combinato disposto dell'art. 5 l. 689/1981 e 213 comma 4 codice stradale? Foro napoletano 2-3:877–879

Suchman LA (1987) Plans and situated actions: the problem of human-machine communication. Cambridge University Press, Cambridge

Suhir E (2018) Human-in-the-loop. Probabilistic modeling of an Aerospace Mission Outcome. CRC Press/Taylor & Francis Group, Boca Raton

Summerton J, Berner B (eds) (2003) Constructing risk and safety in technological practice. Routledge, London

Sumwalt RL (2007) Do you have a safety culture? Flight Safety Foundation, Aero Safety World, pp 38 ff

Sweeney DH (2016) Oil and gas joint operating agreements: a Comparative world-wide analysis. LexisNexis, New York

Taviano M (2018) Il quadro giuridico della just culture nel settore aeronautico italiano e i passi indietro nel d.lgs. n. 173/2017. Rivista di diritto dell'economia, dei trasporti e dell'ambiente, Giureta XVI:245–267

Temple Lang J (1986) Article 5 of the EEC Treaty: the emergence of constitutional principles in the case law of the court of justice. Fordham Int Law J 10(3):503–535

Ter Kulle A (2004) Safety versus justice. CANSO News, 18, pp 1–2

Terrizzi A (2018) Linee guida e saperi scientifici "interferenti": la Cassazione continua a non applicare la legge Gelli-Bianco. Diritto penale contemporaneo 7:93–112

Thew J (2013) No Fear: How 'Just Culture' is an Antidote to Fear-Based Healthcare. http://www.hl7standards.com

Thomas G (2007) A crime against safety. Air Transp World 44:57–5944

Thompson N, Stradling S, Murphy M, O'Neil P (1996) Stress and organizational culture. Br J Soc Work 26(5):647–665

Tirella S (2017) La riforma Gelli è legge. Ecco come cambia la responsabilità medica. Cammino diritto, 28 February 2017

Todeschini N (2017) Approvata la nuova legge sulla responsabilità medica: cosa cambia rispetto alla 'Balduzzi'. Quotidiano giuridico, Ipsoa, Milano

Togan S (2016) The liberalization of transportation services in the EU and Turkey. Oxford University Press, Oxford, p 224 ff

Trögeler M (2010) Criminalisation of aviation accidents: creation of a just culture: balancing aviation safety and the proper administration of justice. Leiden University, Law School, International Institute of Air and Space Law, Leiden

Trögeler M (2011) Criminalisation of air accidents and the creation of a Just Culture. Diritto dei trasporti, pp 1–44

Trovò L (2008) Eurocontrol: in assenza di attività economica non scattano le norme sulla concorrenza. Diritto dei trasporti:163–170

Trovò L (2011) Il processo d'integrazione degli spazi aerei europei: dalla riorganizzazione in blocchi funzionali verso la globalizzazione dell'Air Traffic Management (ATM). Rivista di diritto dell'economia, dei trasporti e dell'ambiente, Giureta:24 ff

Turco Bulgherini E (2009) Le compentenze dell'ENAV in materia di controllo del traffico aereo nel quadro della revisione della parte aeronautica del codice della navigazione. In: Rizzo MP (ed) La gestione del traffico aereo: profili di diritto internazionale, comunitario ed interno, pp 331–372

Turney R (2007) The Überlingen mid-air collision: lessons for the management of control rooms in the process industries. Institution of Chemical Engineers (IChemE) Symposium Series, 153, pp 1–5

Tytgat L (2012) The Single European Sky Regulation in Europe: new scenarios. In: Pellegrino F (ed) Air navigation rules and practices in Europe: towards harmonization. Giuffré, Milano, pp 93–96

Udvarhelyi B (2016) Supranationality versus National Sovereignty – Linking points between EU law and national criminal law. In: MultiScience – XXX microCAD International Multidisciplinary Scientific Conference University of Miskolc, Hungary, 21–22 April 2016

Ulfvengren P, Corrigan S (2015) Development and implementation of a safety management system in a Lean Airline. Cogn Technol Work 17(2):219–236

Uttal B (1983) The corporate culture vultures. Fortune Magazine, 17 October 1983

Vagliasindi GM (2017) The EU environmental crime directive. In: Farmer A, Faure M, Vagliasindi GM (eds) Environmental crime in Europe. Bloomsbury, London, part I, chapter 3

Valente P (2005) Single European Sky: cielo unico europeo. In: Antonini A, Franchi B (eds) Diritto aeronautico a cent'anni dal primo volo. Giuffré, Milano, pp 203–206

Van Antwerpen N (2009) Cross-border provision of air navigation services with specific reference to Europe. Safeguarding transparent lines of responsability and liability. Kluwer, Alphen aan den Rijn

van Asselt MBA, Renn O (2011) Risk governance. J Risk Res 14(4):431–439

van Beuzekom M (2012) Akerboom SP, Hudson PTW, Patient safety in the operating room: an intervention study on latent risk factors. BMC Surg 12(1):10 ff

van Beuzekom M, Akerboom SP, Hudson PTW (2010) Patient safety: latent risk factors. Br J Anaestesia 105(1):52–59

van het Kaar D (2010) Just culture in civil aviation, not just a fait accompli. Journaal Luchrecht, pp 64 ff

Vaughan D (1999) The dark side of organizations: mistake, misconduct, and disaster. Annu Rev Social 25:271–305

Velliscig L (2018) Assicurazione e "autoassicurazione" nella gestione dei rischi sanitari. Giuffré, Milano

Venturato B (2017) In tema di omicidio colposo, nesso di causalità, colpa. Rivista italiana di medicina legale 4:1596–1610

Vernizzi S (2017) Gestore aeroportuale e bird strike, in particolare, il caso Antonov: tutti colpevoli. No, anzi, tutti assolti! In: Busti S, Signorini E, Simoncini GR (eds) L'impresa aeroportuale a dieci anni dalla riforma del codice della navigazione: stato dell'arte. Giappichelli, Torino, pp 255–268

Vicoli D (2003) Scelte del pubblico ministero nella trattazione delle notizie di reato e art. 112 cost.: un tentativo di razionalizzazione. Rivista italiana di diritto e procedura penale, pp 251–293

Viganò F (2006) Problemi vecchi e nuovi in tema di responsabilità penale per medical malpractice. Corriere del merito 2(8):961–976

Viganò F (2009) Riflessioni sulla c.d. "causalità omissiva" in materia di responsabilità medica. Rivista italiana diritto e procedura penale 4:1679–1725

Viganò F (2013) Il rapporto di causalità nella giurisprudenza penale a dieci anni dalla sentenza Franzese, Relazione all'incontro dibattito presso la Corte di Cassazione il 28 novembre 2012. Diritto penale contemporaneo, n. 3

Vincenzi L (2010) L'Unione europea rafforza il cielo unico. Diritto marittimo 1–2:316–319

Vitale G (2018) Diritto processuale nazionale e diritto dell'Unione europea. L'autonoamia procedurale degli Stati membri in settori a dierso livello di "europeizzazione", ed.it, Catania

Von Buri M (1873) Ueber Causalität und deren Verantwortung, Leipzig

Wade M et al (2008) When the line is crossed... Paths to control and sanction behaviour necessitating a state reaction. Eur J Crim Policy Res 14:101–122

Weber L (2017) The Chicago Convention. In: Dempsey PS, Jakhu RS (eds) Routledge handbook of public aviation law. Routledge/Taylor & Francis, Abingdon/New York

Weick KE (1987) Organizational culture as a source of high reliability. Calif Manag Rev 29:112–127

Weick KE, Roberts KH (1993) Collective mind in organizations: heedful interrelating on flight decks. Adm Sci Q 38(3):357–381

Weiner EL, Kanki BG, Helmreich RL (eds) (1993) Cockpit resource management. Academic Press, San Diego

Weston I (2014) Benefits of an aviation just culture. In: Pellegrino F (ed) Legislation and regulation of risk management in aviation activity, vol I. Giuffré, Milano, pp 67–72

Westrum R (1993) Cultures with requisite imagination. In: Wise J, Stager P, Hopkin J (eds) Verification and validation in complex man-machine systems. Springer, New York, pp 401–416

Westrum R (1996) Safety of a technological system. In: NTSB Symposium on Corporate Culture and Transportation Safety Proceedings, 24–25 April 1996, Crystal City, Virginia

Westrum R, Adamski AJ (1999) Organizational factors associated with safety and mission success in aviation environments. In: Garland DJ, Wise JA, Hopkin VD (eds) Handbook of aviation human factors. Lawrence Erlbaum Associates Publishers, Mahwah, pp 67–104

Whitehead D, Welch Dittman P, McNulty D (2017) Leadership and the advanced practice nurse: the future of a changing health-care environment. FA Davis Company, Philadelphia

Wickens CD, Gordon S, Liu Y (1998) An introduction to human factors engineering. Addison Wesley Longman, New York

Wickens CD, Hollands JG, Banbury S, Parasuraman R (2016) Engineering psychology and human performance, 4th edn. Routledge/Taylor & Francis, London/New York

Wiegmann DA, Shappell SA (2003) A human error approach to aviation accident analysis: the human factors analysis and classification system. Ashgate Publishing Ltd., Aldershot

Wiegmann DA, Zhang H, von Thaden T, Sharma G, Mitchell A (2002) A synthesis of safety culture and safety climate research. Aviation Research Laboratory, University of Illinois, Urbana-Champaign

Wiegmann DA, Zhang H, von Thaden TL, Sharma G, Gibbons AM (2004) Safety culture: an integrative review. Int J Aviat Psychol 14(2):117–134

Winger B, Karner E, Oliphant K (eds) (2018) Essential cases on misconduct. Digest of European tort law, vol 3. The Gruyter, Vienna

Wise J, Stager P, Hopkin J (eds) (1993) Verification and validation in complex man-machine systems. Springer, New York

Woods CP (2017) Operationalizing a just culture policy using a just culture decision guide and toolkit: a DNP. Southeastern Louisiana University, School of Nursing, Hammond

Woods M, Woods MB (2008) Air disasters. Lerner, Minneapolis

Woods D, Johannesen LJ, Cook RI, Sarter NB (1994) Behind human error: cognitive systems, computers, and hindsight. Crew Systems Ergonomics Information Analysis Center, Wright-Patterson Air Force Base

Woodward S (2017) Rethinking patient safety. Boca Raton, Florida

Xerri A (2012) Cielo unico europeo: riflessioni su un diritto aeronautico europeo. In: Pellegrino F (ed) Air navigation rules and practices in Europe: towards harmonization. Giuffrè, Milano, p 67 ff

Yang YZ, Wu LY, Zhang Q (2005) Application of man machine environment system engineering in underground transportation safety. Ind Saf Environ Prot:49–51

Youngberg BJ (2010) Principles of risk management and patient safety. Jones and Bartlett Learning, Sudbury

Youngberg BJ (2012) Patient safety handbook. Jones and Bartlett Learning, Burlington

Youngberg BJ (ed) (2013) Patient safety handbook. Jones & Bartlett Learning, Burlington

Yule S (2003) Safety culture and safety climate: a review of the literature. Industrial Psychology Research Centre, University of Aberdeen, King's College, Aberdeen

Zampone A (1997) La condotta temeraria e consapevole nel diritto uniforme dei trasporti: gli elementi caratterizzanti. In: Studi in onore di G. Romanelli. Giuffré, Milano, pp 1287–1324

Zampone A (1999) La condotta temeraria e consapevole nel diritto uniforme dei trasporti. Giuffrè, Milano

Zampone A (2001) Sulla nozione di wilful nisconduct nella giurisprudenza statunitense alla luce dell'entrata in vigore del Protocollo di Montreal n. 4. Diritto dei trasporti 1:215–219

Zana M (1987) La responsabilità del medico. Rivista critica di diritto privato, pp 159–162

Zana M (2016) Il doppio binario della responsabilità, tra contrattualità ed extracontrattualità. Le novità del disegno di legge Gelli. Le corti fiorentine, pp 19–25

Ziccardi G (2011) Cyber law in Italy. Kluwer, Alphen aan den Rijn

Printed by Printforce, the Netherlands